FM 3-18
SPECIAL FORCES OPERATIONS

MAY 2014

DISTRIBUTION RESTRICTION:

Distribution authorized to U.S. Government agencies and their contractors only to protect technical or operational information from automatic dissemination under the International Exchange Program or by other means. This determination was made on 13 May 2014. Other requests for this document must be referred to Commander, United States Army John F. Kennedy Special Warfare Center and School, ATTN: AOJK-CDI-SFD, 3004 Ardennes Street, Stop A, Fort Bragg, NC 28310-9610.

DESTRUCTION NOTICE:

DESTRUCTION NOTICE. Destroy by any method that will prevent disclosure of contents or reconstruction of the document.

FOREIGN DISCLOSURE RESTRICTION (FD 6):

This publication has been reviewed by the product developers in coordination with the United States Army John F. Kennedy Special Warfare Center and School foreign disclosure authority. This product is releasable to students from foreign countries on a case-by-case basis only.

HEADQUARTERS, DEPARTMENT OF THE ARMY

This publication is available at Army Knowledge Online
(https://armypubs.us.army.mil/doctrine/index.html).
To receive publishing updates, please subscribe at
http://www.apd.army.mil/AdminPubs/new_subscribe.asp.

*FM 3-18

Field Manual
No. 3-18

Headquarters
Department of the Army
Washington, DC, 28 May 2014

Special Forces Operations

Contents

		Page
	PREFACE	iv
	INTRODUCTION	v
Chapter 1	THE HISTORY OF UNITED STATES ARMY SPECIAL FORCES	1-1
	World War II	1-1
	Cold War	1-3
	Modern Special Forces	1-8
Chapter 2	THE ROLE OF SPECIAL FORCES	2-1
	Introduction	2-1
	Strategic Context	2-2
	Special Forces Operations Within the Range of Military Operations	2-5
	Army Unified Land Operations	2-6
	Special Forces–Conventional Force Coordination and Integration	2-6
	The Nature and Limitations of Special Forces	2-8
Chapter 3	GUIDANCE AND PRINCIPAL TASKS	3-1
	Guidance to Conduct Special Forces Operations	3-1
	Special Forces Principal Tasks	3-3
Chapter 4	ORGANIZATION	4-1
	United States Special Operations Command	4-1

DISTRIBUTION RESTRICTION. Distribution authorized to U.S. Government agencies and their contractors only to protect technical or operational information from automatic dissemination under the International Exchange Program or by other means. This determination was made on 13 May 2014. Other requests for this document must be referred to Commander, United States Army John F. Kennedy Special Warfare Center and School, ATTN: AOJK-CDI-SFD, 3004 Ardennes Street, Stop A, Fort Bragg, NC 28310-9610.

DESTRUCTION NOTICE. Destroy by any method that will prevent disclosure of contents or reconstruction of the document.

FOREIGN DISCLOSURE RESTRICTION (FD 6): This publication has been reviewed by the product developers in coordination with the United States Army John F. Kennedy Special Warfare Center and School foreign disclosure authority. This product is releasable to students from foreign countries on a case-by-case basis only.

*This publication supersedes FM 3-18, 5 March 2012.

Contents

	United States Army Special Operations Command	4-1
	United States Army John F. Kennedy Special Warfare Center and School	4-2
	United States Army Special Forces Command	4-2
Chapter 5	**EMPLOYMENT**	**5-1**
	Country Team	5-1
	Theater of Operations Organization	5-3
	Training Programs	5-23
Chapter 6	**PLANNING CONSIDERATIONS FOR UNCONVENTIONAL WARFARE AND FOREIGN INTERNAL DEFENSE**	**6-1**
	Unconventional Warfare Considerations	6-1
	Foreign Internal Defense Considerations	6-10
Chapter 7	**SUSTAINMENT**	**7-1**
	Army Special Operations Forces Support to Special Forces Operations	7-1
	Host-Nation Support	7-2
	Contractor Support	7-3
	Statement of Requirements	7-5
	Planning and Executing Theater Operations Support	7-6
	Army Sustainment Principles	7-6
	SOURCE NOTES	**Source Notes-1**
	GLOSSARY	**Glossary-1**
	REFERENCES	**References-1**
	INDEX	**Index-1**

Figures

Figure 2-1. Role of Army special operations forces 2-7
Figure 3-1. Special Forces principal tasks 3-4
Figure 3-2. Role of Special Forces in unconventional warfare and foreign internal defense 3-8
Figure 4-1. United States Special Operations Command organization 4-1
Figure 4-2. United States Army Special Operations Command organization 4-2
Figure 4-3. United States Army Special Forces Command (Airborne) organization 4-3
Figure 4-4. Regional orientation of the Special Forces groups 4-4
Figure 4-5. Special Forces group (airborne) organization 4-5
Figure 4-6. Headquarters and headquarters company (airborne) organization 4-7
Figure 4-7. Group support battalion (airborne) organization 4-8
Figure 4-8. Special Forces battalion (airborne) organization 4-11
Figure 4-9. Battalion headquarters and headquarters detachment (airborne) organization 4-13
Figure 4-10. Battalion support company (airborne) organization 4-14
Figure 4-11. Special Forces company (airborne) organization 4-16
Figure 4-12. Special Forces operational detachment—bravo organization 4-18

Figure 4-13. Special Forces operational detachment—alpha organization 4-19
Figure 5-1. Country team concept 5-2
Figure 5-2. Theater command structure 5-3
Figure 5-3. Theater and Korean Peninsula special operations commands 5-4
Figure 5-4. Example of distributive command and control 5-6
Figure 5-5. Notional joint task force command and control 5-8
Figure 5-6: Notional theater special operations joint task force 5-10
Figure 5-7. Notional joint special operations task force 5-12
Figure 5-8. Example of Combined Forces Special Operations Component
  Command—Afghanistan organization 5-13
Figure 5-9. Notional special operations task force organization 5-15
Figure 5-10. Operations center organization 5-16
Figure 5-11. Notional operations center 5-17
Figure 5-12. Signal center organization 5-19
Figure 5-13. Support center organization 5-20
Figure 5-14. Advanced operations base organization 5-21
Figure 5-15. Example of special operations command and control
  element organization 5-23
Figure 6-1. Advanced operations base providing mission command
  systems, logistics, and advisory assistance 6-15
Figure 6-2. Special Forces element relationship during the conduct of theater
  security cooperation plan events in the steady state 6-16
Figure 6-3. Advanced operations base and special operations task force
  providing advisory assistance 6-17

Tables

Table 4-1. Special Forces core attributes 4-23
Table 4-2. Adaptability competencies 4-24
Table 4-3. Advanced skills on a Special Forces operational detachment—alpha 4-29
Table 7-1. Medical personnel structure within the Special Forces group 7-15

Preface

FM 3-18 is the principal manual for Special Forces (SF) doctrine. It describes SF roles, missions, capabilities, organization, mission command, employment, and sustainment operations across the range of military operations. This manual is a continuation of the doctrine established in the JP 3-05 series, ADP 3-05, ADRP 3-05, and FM 3-05.

The principal audience for FM 3-18 is all members of the profession of arms. Commanders and staffs of Army headquarters serving as joint task force (JTF) or multinational headquarters should also refer to applicable joint or multinational doctrine concerning the range of military operations and joint or multinational forces. Trainers and educators throughout the Army will also use this publication.

Commanders, staffs, and subordinates ensure their decisions and actions comply with applicable United States, international, and, in some cases, host-nation laws and regulations. Commanders at all levels ensure their Soldiers operate in accordance with the law of war and the rules of engagement. (See FM 27-10.)

FM 3-18 uses joint terms where applicable. Selected joint and Army terms and definitions appear in both the glossary and the text.

FM 3-18 applies to the Active Army, the Army National Guard (ARNG)/Army National Guard of the United States, and the United States Army Reserve (USAR) unless otherwise stated.

The proponent of FM 3-18 is the Special Operations Center of Excellence. The preparing agency is the SF Directorate, United States Army John F. Kennedy Special Warfare Center and School. Send comments and recommendations on a DA Form 2028 (Recommended Changes to Publications and Blank Forms) to Commander, USAJFKSWCS, ATTN: AOJK-CDI-SFD, 3004 Ardennes Street, Stop A, Fort Bragg, North Carolina 28310-9610; by e-mail to AOJK-DT-SF@soc.mil; or by electronic DA Form 2028. Unless this publication states otherwise, masculine nouns and pronouns do not refer exclusively to men.

ACKNOWLEDGMENT

All images are courtesy of the United States Special Operations Command (USSOCOM).

Introduction

The United States (U.S.) Army SF represents the premier Department of Defense (DOD) force designed to conduct special warfare. SF Soldiers are specifically selected, trained, and educated to shape foreign political and military environments by working with and through host nations (HNs), regional partners, and indigenous populations. SF provides the United States with a small-footprint option for influencing unfriendly regimes, addressing insurgencies, and containing conflicts that could destabilize U.S. allies and partners. To accomplish these missions, SF possesses qualities and capabilities to mix nonlethal and lethal activities designed to shape the environment, deter conflict, prevail in war, or successfully conduct a wide range of contingency operations.

The two primary missions of SF are unconventional warfare (UW) and foreign internal defense (FID), which can be considered conceptual opposites. In UW, SF works through resistance movements to coerce, disrupt, or overthrow unfriendly governments (or occupying powers) through a wide combination of low-visibility direct and indirect activities in denied areas. In a FID mission, SF works with friendly governments through a wide combination of military engagement activities that enhance the overall security of that nation. Regardless of the mission, the selection, training, and education of the SF Soldier to work with indigenous forces while conducting UW are also applicable to the conduct of FID. In both missions, SF focuses on engaging with and empowering indigenous partners to act. Both efforts are population centric and can be used to set the conditions needed for successful unified land operations.

FID activities provide the United States with a means to strengthen our understanding of regional issues and, through security cooperation, to prevent conflict or intelligently address ongoing conflict. SF provides an experienced and mature military advisory capability that assists allies and partner nations to develop regional stability, enhance global security, and facilitate future operations. FID is designed to shape the environment and prevent or deter conflict. Shaping facilitates potential future offensive, defensive, and stability tasks, such as humanitarian assistance, counterinsurgency, and counterterrorism activities. Prevention is designed to diminish the creation of adversaries or the initiation of hostilities through security cooperation in support of ongoing diplomatic and developmental programs. Deterrence dissuades adversaries from contemplated violence by the development and training of credible and effective HN security forces, enabling them to defeat terrorists or insurgents in their country before they threaten the United States. Over the long term, persistent engagement with HNs, regional partners, indigenous populations, and their institutions through FID activities help set conditions for best possible outcomes.

UW is the primary mission that most distinguishes SF. The inherently interagency, multinational, and widely dispersed activities of UW are conducted to enable a resistance movement or insurgency by operating through or with an underground, auxiliary, or guerrilla force in denied areas to coerce, disrupt, or overthrow oppressive regimes. UW operations are politically sensitive activities that involve a high degree of military risk and, therefore, require distinct authorities and innovative campaign design. The demands of UW missions require mature SF Soldiers who are adept at interacting with a wide range of actors and agencies capable of, and trusted with, sensitive and largely independent operations. The payoff—what makes SF UW activities worth the risk to forces in a denied area—is the subsequent weak defensive posture the enemy must take in order to defend everything and thereby defend nothing well.

Whether the mission is UW, FID, or simply direct action, SF is designed and trained to understand the balance between diplomacy and force, and which is most applicable for creating the conditions for a favorable conflict resolution. This judgment has been honed over decades of engagement with allies, coalition partners, and indigenous populations and provides the United States a discreet, low-signature, small-footprint alternative to a JTF or other large military force structure.

Chapter 1 discusses the history of the SF regiment and how it evolved into its current form. It also provides valuable insight into its design, characteristics, and capabilities.

Chapter 2 covers the role of SF.

Introduction

Chapter 3 provides detailed information on guidance and principal tasks of SF.

Chapter 4 details the current SF organization.

Chapter 5 outlines SF employment.

Chapter 6 provides planning considerations for UW and FID.

Chapter 7 discusses the unique sustainment requirements of SF.

FM 3-18 provides a starting point for conducting Army SF operations. It establishes a common frame of reference and offers intellectual tools Army leaders use to plan, prepare, execute, and assess SF operations. By establishing a common approach and language for SF, doctrine promotes mutual understanding and enhances effectiveness during operations. The doctrine in this publication is a guide for action rather than a set of fixed rules. In Army SF operations, effective leaders recognize when and where doctrine, training, or even their experience no longer fits the situation, and adapt accordingly.

FM 3-18 does not add or modify any terminology found in the Army lexicon and is not the source document for any terms.

Chapter 1

The History of United States Army Special Forces

In any campaign in the future, the participant who fails to take proper cognizance of guerrilla warfare, either in the planning or execution stages, may suffer severe setbacks, if not eventual defeat.

Colonel Russell W. Volckmann
September 1950

There has always been a romantic fascination with special operations forces (SOF). The idea of secret commandos or Rangers striking from the shadows surprising the enemy with overwhelming speed, violence of action, and cutting-edge technology appeals to America's image of highly trained, elite Soldiers. There is, however, another Soldier who fights from the shadows. This one is perhaps less known and far less understood. His real weapons are a deep understanding of terrain, the relationships built, and the influence developed to motivate and train others to take up the fight. These Soldiers are the U.S. Army SF, the "quiet professionals" whom history and popular culture often overlook. Designed to organize, train, and support indigenous personnel in behind-the-lines resistance activities, SF belongs to an organization unique in the Army's history. Founded at the Psychological Warfare Center at Fort Bragg in 1952 and based upon lessons learned and formation used in guerrilla warfare during World War II, its sole purpose was UW. The experience in Vietnam gave SF a second purpose: countering a subversive insurgency. This brief history identifies the precursors and major developments that created modern U.S. Army SF.

WORLD WAR II

1-1. In order to understand a thing one should understand its original purpose—its reason for being. Whereas special operations and irregular forces can trace their origins as far back as the foundation of our nation, the genesis of SF began in the early days of World War II. Following the German invasion of Western Europe in June 1940, thousands of people in France, Belgium, Holland, Denmark, and Norway refused to accept defeat and sought to resist the Axis occupation. To exploit these potential forces behind the German lines, the British organized a special unit called the Special Operations Executive. This unit was shrouded in secrecy and known by such innocuous names as the Joint Technical Board or the Inter-Service Research Bureau in order to conceal its purpose. Unit members and those "in the know" used another name—the Ministry of Ungentlemanly Warfare—which revealed its true nature.

1-2. The broad mission of the Special Operations Executive was to attack German war potential wherever it was exposed, thereby causing a drain on German resources. The Executive was also to organize guerrilla forces that would give maximum assistance to the forces of liberation when the continent was ultimately invaded.

OFFICE OF STRATEGIC SERVICES

1-3. In the beginning of World War II, America's intelligence organizations operated independently of one another and lacked coordination and cooperation. The Army, Navy, Federal Bureau of Investigation, and Department of State all ran their own separate intelligence operations. In 1941, with war looming on the horizon, Colonel William J. Donovan—a World War I Medal of Honor recipient—was named the head of the Coordinator of Information, America's first national intelligence agency. Donovan fought to bring together various activities and functions under one umbrella organization. In 1942, this organization was renamed the Office of Strategic Services.

1-4. Before America entered the war, Donovan personally studied the Special Operations Executive. He admired the organization in coordinating the Political Warfare Executive intelligence activities with psychological warfare and special operations. To Donovan, intelligence penetration provided the basis for planning and propaganda, which would be followed by special operations involving sabotage, subversion, commando raids, guerrilla actions, and behind-the-lines resistance movements. Donovan would use the Special Operations Executive model to organize the Office of Strategic Services into separate but supporting functions of intelligence, propaganda, and guerrilla warfare.

1-5. Subordinate to the newly formed Joint Chiefs of Staff, the Office of Strategic Services drew on the military Services for most of its personnel. Chartered under Joint Chiefs of Staff Directive 155/11/D, the Office of Strategic Services was assigned the responsibility to conduct secret intelligence, research and analysis, and secret operations and planning. Secret operations consisted of morale subversion (psychological warfare) and physical subversion or special operations. The Office of Strategic Services was divided into a number of branches, with each branch being responsible for a particular function. Under the strategic services operations function the Special Operations Branch was designed to organize and support sabotage operations behind enemy lines. They furnished agents, communications, and supplies to underground and guerrilla groups in Norway, France, Denmark, Italy, and China. Special Operations Branch was sub-organized into special teams designed to work behind the enemy lines for the destruction of specific targets, the securing of intelligence, and conduct of guerrilla warfare. Working alongside the British, the Office of Strategic Services/Special Operations Executive combined special operations took on a new name: Special Forces.

1-6. While the new SF name was used at the headquarters (Special Forces Headquarters) and for liaisons to various commands (Special Forces Detachments) the Office of Strategic Services special operations elements remained known by their specific organizational names—the operational groups and the Jedburghs.

Operational Groups

1-7. Of the two Office of Strategic Services special operations elements, the operational groups most resembled the modern SF operational detachments. Designated as 2671st Special Reconnaissance Battalion, Separate (Provisional), in 1944, the operational groups were designed and organized to stand-up and train guerrilla forces. They received specialized instruction in language, foreign weapons, operation and repair of indigenous vehicles, communications, demolitions, organizing and training of civilians for guerrilla warfare, and airborne and amphibious operations. These small teams worked in uniform and could employ as a headquarters element and two sections comprised of 2 officers and 13 enlisted men, similar to the organization Colonel Aaron Bank—the first commander of a Special Forces group (SFG)—would use eight years later.

1-8. The operational groups were the largest Office of Strategic Services effort in France, involving 356 Americans. Prior to and immediately after D-day, the Operational Group Command parachuted teams behind enemy lines to provide liaison and support to the French resistance, the Maquis. The operational groups with the Maquis harassed enemy lines of communications, attacked vital enemy installations, organized and trained the local underground, and furnished intelligence. Operational groups also operated in Italy, Greece, Yugoslavia, Burma, Malaya, and China, training and leading indigenous guerrilla forces.

Jedburghs

1-9. The other Office of Strategic Services units similar in nature to modern SF were the Jedburgh Teams, also referred to as Jeds. Their mission, Operation JEDBURGH, represented the first real cooperation in Europe between Special Operations Executive and Office of Strategic Services Special Operations Branch. Taking their name from the town of Jedburgh in southeastern Scotland (selected at random from a list of pre-approved code names), these three-man teams were a mixture of Office of Strategic Services Special Operations Branch, Special Operations Executive, and Free French Bureau

Central de Renseignements et d'Action personnel. The Jedburgh teams consisted of three men: a commander, an executive officer, and a noncommissioned officer radio operator. One of the officers would be British or American, while the other would originate from the country to which the team deployed.

1-10. Like the operational groups, the Jedburgh's mission in France was to support and coordinate the French resistance Maquis activities under the sponsorship and guidance of the Supreme Headquarters Allied Expeditionary Force. As a secondary role, Jedburgh teams engaged in guerrilla warfare and attacks on German lines of communications. Jedburghs also operated under the South East Asia Command in 1945, including French Indochina and Burma, working with antifascists and Karen tribesmen. If the operational groups provided the organizational concept for SF, the Jedburghs provided many of the original personnel. Many former Jedburghs would go on to join SF, including its first commander, Colonel Aaron Bank.

Detachment 101

1-11. The best-known and most successful Office of Strategic Services guerrilla operation in the war is attributed to Detachment 101 in Burma, commanded by Colonel William R. Peers. Detachment 101 organized and trained Kachin tribesmen to conduct guerrilla operations against the Japanese in 1943 and 1944 and represented a microcosm of the entire range of Office of Strategic Services capabilities including the gathering of intelligence, aiding escape and evasion of downed fliers, espionage, counterespionage, sabotaging lines of communications and activities in support of allied conventional operations. Detachment 101 received the Distinguished Unit Citation from President Eisenhower for its service in the 1945 offensive that liberated Rangoon.

GUERRILLAS WARFARE IN THE PHILIPPINES

1-12. Whereas the personnel and organizational concept for SF primarily came from the Office of Strategic Services operations in Europe, much of the doctrine came from the Philippines. After the fall of Bataan in 1942 to the invading Japanese, several U.S. Army officers serving in the Philippines refused to surrender to the Japanese and organized resistance from the local forces and local population.

1-13. Captain Russell Volckmann, Captain Donald Blackburn, and Colonel Wendell Fertig led Filipino guerrillas in harassing the Japanese. Upon General MacArthur's return to the Philippine Islands, they continued their guerrilla warfare campaign by attacking retreating Japanese forces, capturing bases and airfields, and helping speed the American advance. Volckmann's guerrillas collaborated with the 6th Ranger Battalion and the Alamo Scouts and supported their missions in Luzon, including the rescue of the prisoners at Cabanatuan. On Mindanao, Fertig led guerrilla forces in hit-and-run attacks and coast-watching activities. It is estimated that Fertig commanded the equivalent of an Army corps of 36,000 guerrillas, holding down up to 60, 000 Japanese forces and killing at least 7,000. After the war Fertig and Volckmann would add their experience joining the former Office of Strategic Services members in standing up the America's first dedicated guerrilla warfare unit, U.S. Army SF.

COLD WAR

1-14. As World War II ended, so did the perceived need for a full-time organization dedicated to special operations and intelligence. Executive Order 9621, issued on 20 September 1945, provided only 10 days to dissolve the Office of Strategic Services, with the Research and Analysis Sections going to the State Department and the War Department receiving the remainder for salvage and liquidation. Fortunately, the Assistant Secretary of War, John McCoy, saved the Secret Intelligence (commonly referred to as SI) and Counterespionage (also called X-2) Branches, which became the nucleus of the Central Intelligence Agency under the National Security Act of 1947. While the fledgling Central Intelligence Agency assumed the intelligence functions of the Office of Strategic Services, the Office of Policy Coordination was stood-up to manage the support to resistance and underground networks previously developed by the Office of Strategic Services.

1-15. In 1949, the head of Office of Policy Coordination, Frank Wisner, requested assistance from the Army in the training of personnel for guerrilla warfare. Although the peacetime mission belonged to the

Office of Policy Coordination, much of the guerrilla warfare expertise still resided in the Army. Training in guerrilla warfare was established at the Infantry School in Fort Benning, Georgia. Shortly after, Special Text 31-90-1, *Operations Against Guerrilla Forces*, written by (then) Lieutenant Colonel Russell Volckmann, was published. This document, the U.S. Army's first dedicated guerrilla warfare doctrine supported the notion that the master of UW is the best resource for counterguerrilla operations. In 1951, Volckmann also wrote the first FM on guerrilla warfare—FM 31-21, *Organization and Conduct of Guerilla Warfare* [sic], which used the organizational concept of an operational group to refer to the small unit linking up with guerrilla forces. Its mission was to organize existing guerrilla forces or resistance groups capable of conducting strategic and tactical operations against an enemy. The theater element charged with managing and coordinating special operations within the theater command was called the Special Forces element.

1-16. After World War II, Army Chief of Staff Dwight Eisenhower directed that a psychological warfare capability using experienced personnel be maintained. Principal among those with experience was Brigadier General Robert McClure, who headed Eisenhower's psychological warfare effort in the European theater during the war. Like Donovan, McClure believed in the linkage between morale operations (psychological warfare operations) and sponsoring resistance forces.

1-17. The outbreak of the Korean conflict heightened interest and the Psychological Warfare Division was established in September 1950. Shortly after, this organization was renamed the Office of the Chief of Psychological Warfare, with Brigadier General McClure working directly with the Army Chief of Staff. Similar to Donovan at the Office of Strategic Services, McClure organized the Office of the Chief of Psychological Warfare into Propaganda, UW, and Support Divisions. Army Special Regulation 10-250-1, *Organization and Functions, Department of the Army, Office of the Chief of Psychological Warfare, Special Staff* (22 May 1951), gave McClure the authorization to formulate and develop psychological and special operations plans for the Army and supervise Army programs in this field.

1-18. In May 1952, the Psychological Warfare Center at Fort Bragg, North Carolina, was established. The Office of the Chief of Psychological Warfare Special Operations Division sought out former Office of Strategic Services officers (such as Bank) and others who had experience with guerrilla warfare—men like Fertig and Volckmann. Volckmann was working on joint Army-Central Intelligence Agency operations behind the lines in North Korea. Psychological warfare via McClure provided the platform to sell the Army on the need for a rebirth of the Office of Strategic Services Special Operations Jedburghs and operational groups—now generally referred to as Special Forces units.

1-19. The Korean conflict was a devastating blow to the Army. Ranger companies took up to 50-percent combat losses and were in danger of being disbanded. McClure's Office of the Chief of Psychological Warfare authored a study titled *Special Forces-Ranger Units* that outlined the concept for their proposed special operations-type operational unit. The new organization was called Special Forces—a designation derived from the Office of Strategic Services whose operational teams in the field were given the same name in 1944. Army Field Forces requested the removal of any reference or tie-in to the Rangers because they "envisioned Special Forces will in all probability be involved in subversive activities" and should focus SF on using indigenous groups. On 23 August 1951, General Maxwell Taylor decided to deactivate all Ranger units, at which point McClure and his staff briefed Taylor on the mission and capabilities of an SF organization. As a result, the personnel spaces needed to create 10th SFG came from those deactivated Ranger companies.

1-20. The concept of a full-time Army unit dedicated to UW had many opponents after World War II. Many considered UW a task for conventional formations with no need for another school and another elite unit, especially during downsizing of the force. The newly formed Air Force, along with many in the newly formed Central Intelligence Agency, believed their joint capabilities could effectively support, organize, and utilize resistance forces. McClure's background and contacts in running psychological operations—not only during war, but also during the post-conflict years—built his credibility with the men who would be Army Chiefs of Staffs for years to come. His deep understanding of the inner workings of the War Department, the Pentagon, and (later) the DOD gave him the ability to find and recruit men like Bank, Volckmann, Blackburn, and Fertig, and set the necessary conditions to be in the right place at the right time to stand up a UW capability in the Army.

1-21. As a result of McClure's efforts, the Army's first SF unit was activated at Smoke Bomb Hill on Fort Bragg, North Carolina, with former Office of Strategic Services Jedburgh Colonel Aaron Bank selected from McClure's staff to be the commander. Present for duty on that day—19 June 1952—were seven enlisted men, one warrant officer, and Colonel Bank. Volunteers were drawn from personnel with guerrilla warfare experience in World War II or Korea, Office of Strategic Services members, and Lodge Bill Soldiers (Eastern European or stateless volunteers in the U.S. Army). The unit designation as the 10th SFG was a nod to its psychological operations background; by starting with the number 10, they hoped to deceive observers into thinking there were at least 9 other SF units.

1-22. First and foremost, the 10th SFG was created in the Office of Strategic Services Special Operations/Special Operations Executive model to stand-up indigenous resistance forces in the eventuality that Communist forces—namely the Soviets and the Warsaw Pact—invaded Europe. The key replacement for the Maquis partners would then be émigrés from Eastern Europe who had the regional and language expertise for the UW mission, recruited under the 1950 Lodge-Philbin Act (commonly known and the Lodge Act).

First Employments

1-23. During the Korean conflict, when the 8th Army retreated from the Yalu in 1950, as many as 10,000 Korean irregulars declared their willingness to fight with the United States against the Communist forces. Most of these partisans fled to friendly-held islands off the Korean coast and formed the United Nations Partisan Forces, Korea. Volckmann had been reassigned from Fort Benning, where he was developing counterguerrilla training, to Korea in order to work with these partisan and plan behind-the-lines operations. The partisans were trained, supported, and directed by the Army, Air Force, and Central Intelligence Agency. In 1951, the partisans—organized as the 8240th Army Unit—were given the mission to establish a resistance net in North Korea in anticipation of a United Nations offensive to liberate all or part of North Korea—an offensive that never did occur. The primary United Nations objectives were to maintain the status quo and negotiate a peace. As a result, the original covert mission gave way to small-scale harassment against the enemy coastal flanks. The force strength of the partisans peaked at approximately 23,000 before they eventually were inducted into the Republic of Korea Army. Partisan forces reported 4,445 actions, resulting in 69,000 enemy casualties. This was the U.S. Army's first experience employing partisans against a Communist enemy, using some 200 U.S. Army personnel, and would represent the first chance for the fledgling SF to conduct its designated mission. Very few SF Soldiers made it to Korea, and most of those were individual replacements for conventional forces. Although they contributed to the guerrilla warfare effort, their participation would not prove to be the event that validated the new concept.

1-24. The first unit mission came as a result of an aborted uprising in East Germany on 11 November 1953. The possibility of capitalizing on this resistance to Communist occupation led to the decision to deploy half of the 10th SFG to Bad Tolz, West Germany. Those SF Soldiers still in training at Fort Bragg who did not deploy remained under the command of former Philippine guerrilla leader (then) Colonel Donald Blackburn and were redesigned as the 77th SFG. As detachments became qualified and operational, they were deployed to pre-stage in Thailand, Taiwan, Japan, Hawaii, and Vietnam. Several of these detachments later relocated to Okinawa and formed the nucleus of the 1st SFG, activated in 1957. The 77th SFG at Fort Bragg would later be redesignated the 7th SFG.

1-25. These original operational detachments were designed around the Office of Strategic Services operational group construct and it was understood that they were independent or "detached" for operations. The original designation for the operational element called the force area team A (FA) team came from Volckmann's 1951 FM 31-21, which referred to the forward element under SF as Guerrilla Force Area A. Two years later, an FA team (equivalent to the modern SF operational detachment—alpha [ODA]) consisted of 15 men and was designed to advise a regiment of 1,500 guerrillas. A force area team B (FB) team (equivalent to a modern SF operational detachment—bravo [ODB] or SF company headquarters) was also principally a guerrilla advisory element, but could also command two or more FA teams. A force area team C (FC) team (SF operational detachment C or SF battalion headquarters equivalent) was the next level of advisory and could also command FA and FB teams. The force area team D (FD) team, called the group headquarters, could command the entire group and manage guerrilla campaigns in two or more countries.

1-26. All these elements were to deploy under the command and control of a theater SF command later designated the Joint Unconventional Warfare Task Force. Administrative teams (AA and AB teams) were support elements for the FA and FB teams before deployment. Once teams went forward to link up with area guerrilla forces, AA and AB teams melded with the group headquarters. The organization was very flexible when forward. In the rear, it appeared as teams run by captains within companies run by lieutenant colonels, under one group headquarters commanded by a colonel. In the late 1960s and early 1970s, the force was organized into groups, battalions, companies, and teams—similar to a conventional infantry formation.

1-27. The Cold War provided other opportunities for employment of SF units. During August 1956, six SF operational A detachments of the 10th SFG stationed at Flint Kaserne in Bad Tolz, West Germany, were reassigned to West Berlin under the 7761st Army Unit (subsequently known as 39th Special Forces Operational Detachment) and embedded within Headquarters and Headquarters Company (HHC), 6th Infantry Regiment. Their mission was stay-behind UW. In April 1958, the unit found its final home at Andrews Barracks, West Berlin, and was assigned to HHC, U.S. Army Garrison Berlin, with a new name—Detachment A (or Det. A). Later the unit became Detachment A, Berlin Brigade, U.S. Army Europe, which it remained until its deactivation in 1984.

1-28. In 1955, SF received its first publicity—two articles in the *New York Times* announcing the existence of a U.S. Army "liberation force" designed to fight behind enemy lines. The *Times* correspondent noted the distinct foreign nature of the SF, as many of its volunteers were refugees from Eastern Europe. Photographs showed 10th SFG troops wearing berets with their faces blacked out in the photos to conceal their identity. As the 10th SFG became established in Germany, a new item of headgear—the green beret—appeared in rapidly increasing numbers. The group commander, Colonel Eckman, authorized the wear of the beret and it became unit policy in 1954. By 1955, every SF Soldier in Germany was wearing the green beret as a permanent part of his uniform. The Department of the Army, however, did not recognize the headgear.

INDOCHINA

1-29. The Peoples Revolution in China and the Korean Conflict set the stage for the Cold War sweeping across Asia. Stopping the spread of Communism became foreign policy in the region—much like stopping the spread of global terror is today. In 1953, Dwight Eisenhower appointed Office of Strategic Services founder William Donovan as the U.S. Ambassador to Thailand. Brigadier General McClure, who stood up the Psychological Warfare Center at Fort Bragg, was retasked by Army Chief of Staff General J. Lawton Collins to become the chief of the U.S. military mission to Iran. McClure participated in Operation AJAX, considered the Central Intelligence Agency's first successful covert regime change/UW operation. Around the world, pieces were being put into place by the United States to resist Communism and establish stable governments that would serve as allies in the same effort. With the onset of nuclear deterrence, men like Army Chief of Staff General Maxwell Taylor identified the need to fight limited warfare. General Taylor's book, *The Uncertain Trumpet*, was used by (then) Massachusetts Senator, John F. Kennedy, during the 1960 presidential campaign to criticize Eisenhower's defense policies.

1-30. In 1959, 149 members of 77th SFG (redesignated the 7th SFG in 1960) deployed to Laos to enhance, and later replace, the French Military Mission in order to train and advise the Royal Laotian Army in a then-clandestine mission Operation HOTFOOT. This mission marked the first SF deployment to a country to counter an active insurgency. At the same time SF teams worked with the Central Intelligence Agency in northern Laos, which had been acknowledged as a separate state in the 1954 Geneva Accords, to train and advise over 30,000 indigenous Mien, Khmu, and Hmong hill tribesmen in an ongoing Central Intelligence Agency-run UW operation against the communist Pathet Lao and Viet Minh occupiers. A principal objective of North Vietnam was to maintain access into South Vietnam through Laos. The second rotation, commanded by Lieutenant Colonel Arthur "Bull" Simons, brought in 12 more ODAs as mobile training teams in what was then called Operation WHITESTAR. President Kennedy authorized the withdrawal of WHITESTAR Teams in 1962, and the Military Assistance Advisory Group was moved to Thailand. Today, a granite monument located on the way to the John F. Kennedy memorial at the Arlington National Cemetery commemorates the Hmong and other ethnic tribesmen for the military support given during the fight against Communism in Laos.

1-31. In 1960, the U.S. Army assigned an official lineage to SF consisting of various elite special-purpose units (such as Roger's Rangers, World War II Ranger battalions, and 1st Special Service Force). In response, Colonel Bank wrote that "the only precursor of U.S. Army Special Forces is the Office of Strategic Services." In 1986, the Ranger lineage was removed and made part of the lineage of the newly created Ranger battalions. Lineage and honors are established by the Center for Military History in accordance with AR 870-5. When units are scheduled for activation, the Center for Military History reviews all inactive units and selects the one with the best history in an effort to establish continuity, entitle honors and heraldic items, and pass on organizational historical property. The problem with SF's lineage is that the true origin was a civilian organization—the Office of Strategic Services—which could not be considered. Disbanded elite units were the next-best option; therefore, the 1st Special Service Force is the official heritage for 1st Special Forces, along with all its campaign participation credits from World War II. It is for this reason that 1st Special Forces and all the SFGs in the regiment are given credit for Aleutian Islands, Naples-Foggia, Anzio, Rome-Arno, Southern France, and Rhineland. Secondly, and perhaps of greater relevance in this case, the Secretary of the Army may endow honors to an organization when there is no lineal connection in order to entrust an active organization to properly conserve the heritage of a magnificent military unit. This heritage has been entrusted to 1st Special Forces Regiment.

1-32. In 1961, President Kennedy inherited a defense policy based on nuclear deterrence. To meet this challenge, the Air Force chose bomber wings, the Navy chose nuclear submarines and aircraft carriers, and the Army chose large pentomic divisions geared to fight nuclear warfare, which soon after evolved into brigades of armor and mechanized infantry. The Marine Corps was the only ground force that might be suitable for counterguerrilla operations, but they were reluctant to opt for organizing around small units and chose a vertical envelopment concept using rotary-wing aviation. Up to this point, SF had remained small and insignificant. This would all change when President Kennedy decided that a counterinsurgency capability was a critical missing piece in his defense strategy. Under Kennedy's leadership, SF expanded rapidly. The 10th SFG in Bad Tolz, the 1st SFG in Okinawa, and the 7th SFG at Fort Bragg increased fourfold in authorized strength to 1,500 men each. Other groups were planned to cover responsibilities in other parts of the world. Three groups were activated in 1961: the 11th SFG, the 12th SFG, and the 5th SFG, which saw extensive action in Vietnam. Three more groups were activated in 1963—the 6th SFG to cover Southwest Asia, the 8th SFG in the Panama Canal Zone and responsible for Latin America, and the 3d SFG at Fort Bragg to cover Africa and parts of the Middle East. The President also authorized the wear of the green beret as a mark of distinction.

1-33. President Kennedy issued National Security Action Memorandum 124, *Establishment of the Special Group Counter-Insurgency*, in January 1962. National Security Action Memorandum 124 stated that subversive insurgency (what the Soviets called "wars of national liberation") is a major form of politico-military conflict "equal in importance to conventional warfare." The Special Group would ensure that such recognition is reflected in the organization, training, equipment, and doctrine of the U.S. Armed Forces. The specific areas of interest named were Laos, South Vietnam, and Thailand. On 6 June 1962, Kennedy addressed the graduating class at the United States Military Academy at West Point:

> Whatever your position, the scope of your decisions will not be confined to the traditional tenets of military competence and training. You will need to know and understand not only the foreign policy of the United States but the foreign policy of all countries scattered around the world who 20 years ago were the most distant names to us. You will need to give orders in different tongues. . . . You will be involved in economic judgments. . . . In many countries, your posture and performance will provide the local population with the only evidence of what our country is really like. In other countries, your military mission, its advice and action, will play a key role in determining whether those people will remain free. You will need to understand the importance of military power and also the limits of military power, to decide what arms should be used to fight and when they should be used to prevent a fight, to determine what represents our vital interests and what interests are only marginal. Above all, you will have a responsibility to deter war as well as to fight it. For the basic problems facing the world today are not susceptible of a final military solution.

1-34. The Overseas Internal Defense Policy issued by the State Department in September 1962 called on the DOD to develop U.S. military forces trained for employment in UW and counterguerrilla and other military counterinsurgency operations, and to develop language-trained and area-oriented U.S. forces for possible employment in training, advising, or operational support to indigenous security forces. Internal defense is defined at the full range of measures taken by a government to protect its society from subversion, lawlessness, and insurgency.

1-35. After the Kennedy years, the modern era of SF began. Green Berets now had their two principal capabilities—UW and FID (the former by design and the latter by Cold War exigencies). In the years that followed, SF would be called upon to conduct all types of contingency operations, but would stand out most in UW and FID missions. The basic organization, doctrine, recruitment, and training designed by Bank, Fertig, and Volckmann proved to be flexible, durable, and versatile for a broad range of military operations especially against problems that proved resistant to an orthodox military approach.

VIETNAM

1-36. In Vietnam, SF was among the first military units into the theater in the late 1950s and by the early sixties was positioned throughout South Vietnam. SF teams worked as advisors to the Vietnamese Army and the Civilian Irregular Defense Group, trained and led quick-reaction units called Mike Forces, and conducted cross-border operations through such units as the Military Assistance Command Vietnam—Studies and Observation Group. Vietnam became a watershed in the history of SF, spanning over 14 years from 1956, when a 16-member detachment deployed to Vietnam to train a cadre of indigenous Vietnamese SF teams. Throughout the remainder of the 1950s and early 1960s, the number of SF advisors in Vietnam increased steadily. The 5th SFG had been activated at Fort Bragg, North Carolina, to handle the increasing scope and complexity of missions in Vietnam. Signed 28 June 1961, National Security Action Memorandum 57, *Responsibility for Paramilitary Operations*, directed all overt SF paramilitary activities be transferred from the Central Intelligence Agency to the DOD.

1-37. By February 1963, the 5th SFG established a permanent headquarters in Nha Trang, making Vietnam the exclusive operational province of the 5th SFG. SF troops eventually established approximately 254 outposts throughout the country—many of them defended by a single A team and hundreds of friendly natives. Responsible for training thousands of Vietnam's ethnic tribesmen in the techniques of modern warfare, SF Soldiers took the Montagnards, the Nungs, the Cao Dei, and others, and molded them into the 60,000-strong Civil Irregular Defense Group. Other missions included civic-action projects, in which SF Soldiers built schools, hospitals, and government buildings, provided medical care to civilians, and dredged canals. These missions were designed to win the "hearts and minds" of the local population. Gradually, SF turned over its camps to the South Vietnamese, and on 5 March 1971, the 5th SFG returned to Fort Bragg.

1-38. The role of SF in Vietnam was not completely over as some Soldiers continued to serve in various covert missions. In Vietnam, Communist forces depended upon the Ho Chi Minh Trail as their logistical lifeline. Located in Cambodia and Laos, the trail was considered off-limits to U.S. forces. To gather intelligence on supply movements down the trail and to interdict operations, the Military Assistance Command Vietnam—Studies and Observation Group was formed to run clandestine combat operations across the border. The teams from Military Assistance Command Vietnam—Studies and Observation Group were American-led and manned with indigenous forces. In this example of the versatility of SF, these teams and successor units provided needed and timely intelligence for the conventional Army. In recognition of their actions in Vietnam, the men assigned to the 5th SFG received 16 of the 17 Medals of Honor awarded to SF Soldiers in theater (the other was awarded to a Soldier from the 7th SFG).

MODERN SPECIAL FORCES

1-39. After Vietnam, lessons learned were quickly forgotten. The 1st, 3d, 6th, and 8th SFGs were deactivated. Among military leadership, counterinsurgency fell into disrepute. The country was disillusioned with military intervention after a second unsatisfying incursion in Southeast Asia. The Army doctrine once again focused on fighting the modern mechanized forces of the Warsaw Pact. The Army capstone manual—FM 100-5, *Operations* (1976)—focused exclusively on battle in central Europe. Winning wars required superior weapons combined with crews and leaders who could effectively employ them to best effects. In that environment, SF found relevancy by providing reconnaissance in support of conventional forces in major combat operations. Special operations budgets dropped from $1 billion in 1969 to less than $100 million in 1975. Army SF declined from 13,000 string in 1971 to less than 3,000 Green Berets in 1974. Throughout those lean years, the focus of Army SF training at the John F. Kennedy Special Warfare Center changed little since its inception in 1952. The school for special warfare continued to select and train SF personnel to organize, train, and employ a guerrilla force. Military involvements once known as low-intensity conflicts lay ahead and with them came the need for both special warfare capabilities—counterinsurgency and UW.

1-40. During the 1980s, defense received renewed emphasis. 1st SFG was reactivated to cover threats in the Pacific Rim nations. The 3d SFG was reactivated in 1990, reflecting increased concerns in the Caribbean and the African continent. The 7th SFG played a critical role in suppressing the rise of guerrilla movements and containing the spread of Communism in Central and South America by strengthening military capabilities of democratic regimes. This group deployed to El Salvador and developed an effective counterinsurgency force. Soldiers from the 7th SFG also helped the Honduran military resist and defeat an invasion from Nicaragua and a Communist-supported insurgency. During the second half of 1980, the 7th SFG assisted in counternarcotics operations in Venezuela, Colombia, Ecuador, Peru, and Bolivia.

1-41. SF would be institutionally revitalized by way of the 1987 Nunn-Cohen Amendment to the Goldwater-Nichols Act. This legislation effectively placed SOF and low-intensity conflict advocates in the upper levels of the DOD, formalizing the view that SOF were the most appropriate service elements to conduct low-intensity conflict.

1-42. This brief history focuses on the genesis of SF—when Green Berets developed the skill sets needed for special warfare. Those early years created the organization, methods, and mindset that set the stage for a renaissance of UW after 11 September 2001, when operational detachments played a central role in a crushing defeat of our nation's enemies in Afghanistan during Operation ENDURING FREEDOM. UW gained national prominence with images of "horse Soldiers." Operation IRAQI FREEDOM also showcased how small numbers of SF teams, linked with indigenous resistance, could support major combat operations by preventing 13 Iraqi divisions from reinforcing units defending Baghdad.

1-43. When President Kennedy visited Fort Bragg in 1961, SF had grown to over 3,000 Soldiers in three SFGs and the Special Warfare Center and School. Through his endorsement—formal and informal—the President ensured that the U.S. Army SF would be known worldwide as the Green Berets. With President Kennedy's advocacy and direct support, SF grew from three to 11 SFGs by the end of 1963, with seven active SFGs and four additional SFGs in the National Guard and Army Reserve. His words to the SF Commander at Fort Bragg, Brigadier General William P. Yarborough, continue to inspire:

> *I know that you and the members of your command will carry on for us and the free world in a manner which is both worthy and inspiring. I am sure that the Green Beret will be a mark of distinction in the trying times ahead.*

This page intentionally left blank.

Chapter 2
The Role of Special Forces

This is another type of warfare—new in its intensity, ancient in its origin—war by guerrillas, subversives, insurgents, and assassins—war by ambush instead of combat, by infiltration instead of aggression—seeking victory by eroding and exhausting the enemy instead of engaging him. . . . It requires, in those situations where we must counter it if freedom is to be saved, a wholly new kind of strategy, a wholly different kind of force, and therefore a wholly new and different kind of military leadership and training.

President John F. Kennedy, 1962

Characterized by a highly adaptive culture and a versatile organization, SFGs conduct operations that have operational and strategic implications. SF Soldiers operate across the range of military operations, integrating their capabilities with joint, interagency, intergovernmental, and multinational interaction, or with force-multiplying operations with indigenous groups and coalition forces as part of a larger effort. At any given time, SF teams are conducting these types of operations in dozens of countries worldwide.

INTRODUCTION

2-1. SF conducts operations in all operational environments, but is best suited for hostile and uncertain environments. SF can be tailored to achieve not only military objectives through application of SOF capabilities for which there are no broad conventional force requirements, but also to support the application of the diplomatic, informational, and economic instruments of national power. SF operations are typically low visibility or clandestine. SF activities are applicable across the range of military operations. They can be conducted independently or in conjunction with operations of conventional forces, other government agencies, or HNs/partner nations, and may include operations with or through indigenous, insurgent, and/or irregular forces. SF operations differ from conventional operations in degree of political sensitivity and operational risk, dependence on detailed operational intelligence and indigenous assets, operational techniques, and modes of employment.

2-2. The sensitivity of these operations requires a highly mature and culturally intelligent force that is effective in ambiguous, asymmetric, and often austere environments. SF Soldiers are adaptive—a quality based on a selection process that evaluates critical thinking, comfort with ambiguity, acceptance of prudent risk, and the ability to make rapid adjustments based upon a continuous assessment of the situation. The key to SF success is a selection and training program designed to identify and educate Soldiers with the character and attributes necessary to thrive in complex and ambiguous environments. SF training and education programs develop complex problem solvers that can thrive as force multipliers in hostile and unfamiliar environments. Central to the development of the SF Soldier are a deep understanding of cultures and foreign language, proficiency in small-unit tactics, and the ability to fight alongside indigenous combat formations in a permissive or uncertain environment. These skills are built upon continued training in regional orientation and language qualification, skills in team building, cross-cultural communication, and interpersonal relations. While other SOF may share some of the SF principal tasks, the SF role is unique and characterized by its emphasis on UW capabilities. SF Soldiers are routinely entrusted to conduct autonomous operations in remote places and in uncertain—and often hostile—environments. In these circumstances a U.S. Army captain or staff sergeant can create strategic effects as they develop indirect solutions by working with and through indigenous populations

STRATEGIC CONTEXT

2-3. SF operations historically have been used to shape the environment, to conduct condition-setting activities, and to enable maneuver warfare or other operations. The discreet, precise, and scalable nature of SF operations often makes them a more attractive option in instances where a large force structure may be inappropriate or counterproductive or may incur political risk. When used effectively, these types of operations can yield disproportional benefits.

2-4. SF operations are one means by which the President of the United States, Secretary of Defense, Department of State country teams, and geographic combatant commanders (GCCs) can shape an environment to support the U.S. National Security Strategy. The National Security Strategy prepared by the Executive Branch for Congress, outlines the major national security concerns of the United States and how the administration plans to deal with them. It provides a broad strategic context for employing military capabilities in concert with other instruments of national power. SF contributes to the National Security Strategy in the following ways:

- As an instrument to implement or enforce U.S. National Security Strategy outside of an overt military campaign, SF specializes in persistent engagement. SF implements the National Security Strategy developed within a strategic security environment characterized by uncertainty, complexity, rapid change, and persistent conflict. They possess capabilities that enable both lethal and nonlethal missions specifically designed to influence threat, friendly, and neutral audiences. They shape foreign political and military environments by working with HNs, regional partners, and indigenous populations and their institutions. Such proactive shaping can help prevent insurgencies and/or conflicts from destabilizing partners and ultimately deter conflict, prevail in war, or succeed in a wide range of potential contingencies.

- In support of a military campaign, SF acts as a force multiplier through UW and extends the operational reach to influence and strike enemy forces throughout their depth. The use of SF to organize, train, and employ indigenous forces operating in the enemy rear area prevents effective employment of reserves, disrupts command and control and logistics, and forces the enemy to cope with U.S. actions throughout its entire physical, temporal, and organizational depth. Utilizing indigenous information networks provides powerful tools for leaders to synchronize efforts. Synchronized efforts between conventional and unconventional forces offer Army leaders with the capability to integrate actions along with interagency and multinational efforts to overwhelm the enemy and achieve decisive results. UW is the signature excellence of SF. Whether used as a supporting effort to major combat operations or as an alternative, UW strikes the enemy in times, places, and manners for which the enemy is not prepared, seizing the initiative by forcing the enemy to defend everywhere at once.

2-5. Two keys in achieving tactical, operational, and strategic successes are flexibility and adaptability. U.S. Army SF Soldiers work in small teams and are well known for taking initiative, acting quickly, and having an affinity for innovation in thought, plans, and operations—all important factors in a flexible organization. This type of flexibility allows small-footprint, politically sensitive responses when large-scale military employment may be inappropriate or denied. Adaptability first requires an understanding of the operational environment. Sustained engagement around the world, along with scalable features, makes SF teams the most ubiquitous ground forces with both lethal and nonlethal capabilities. Their low profile and an inherent need to network with interagency partners, indigenous populations, and indigenous systems reflect the adaptive nature of SF teams. Their widespread and persistent engagement makes SF teams agile and responsive to rapidly changing regional situations that affect our national security interests.

2-6. The 2010 National Security Strategy reaffirmed the U.S. commitment to retaining its global leadership role and defined U.S. enduring national interests as follows:

- The security of the United States, its citizens, allies, and partners.
- A strong, innovative, and growing U.S. economy in an open international economic system that promotes opportunity and prosperity.

- Respect for universal values at home and around the world.
- An international order advanced by U.S. leadership that promotes peace, security, and opportunity through stronger cooperation to meet global challenges.

2-7. The National Security Strategy and the 2010 Quadrennial Defense Review together guide the establishment of U.S. national military objectives as follows:

- Counter violent extremism.
- Deter and defeat aggression.
- Strengthen international and regional security.
- Shape the future force.

2-8. The 2012 *Sustaining U.S. Global Leadership: Priorities for 21st Century Defense* articulates strategic guidance for the DOD after a decade of war. It shapes a joint force for the future that will be smaller and leaner, but agile, flexible, technologically advanced, and ready to confront and defeat aggression anywhere in the world. It projects a changing security environment of complex challenges and opportunities. Basic tenets include the following:

- Rebalance engagement toward the Asia-Pacific region while continuing defense efforts in the Middle East.
- Conduct counterterrorism and irregular warfare.
- Deter and defeat aggression in one region while committed to large-scale operations elsewhere.
- Project power despite anti-access and area-denial challenges.
- Counter weapons of mass destruction.
- Operate effectively in cyberspace and space.
- Maintain a nuclear deterrent.
- Defend the homeland and provide support to civil authorities.
- Provide a stabilizing presence abroad during a significant reduction in resources.
- Conduct limited stability and counterinsurgency operations.
- Conduct humanitarian, disaster relief, and other operations.

2-9. SF plays a vital role in supporting U.S. national strategy. Through varying applications of UW, FID, counterinsurgency, preparation of the environment, security force assistance, direct action, special reconnaissance, counterterrorism, and counterproliferation of weapons of mass destruction, SF has proven its utility in conflict, cold war, and contingency operations. SF discreetly shapes the operational environment in both peace and complex uncertainty. SF strengthens U.S. interests by sustained engagement with allies and partners.

2-10. SF units normally conduct special operations supporting the theater special operations command (TSOC) within the GCC's area of responsibility. These operations are conducted in support of the U.S. Ambassador and country team or in conjunction with joint operations being conducted in accordance with a command relationship established by the designated joint force commander (JFC). In either situation, SF offers military options for situations requiring subtle, indirect, or low-visibility applications. The small size and unique capabilities of SF give the United States a variety of appropriate military responses. These responses typically do not entail the same degree of political sensitivity or risk of escalation normally associated with the employment of a larger and more visible force.

LEVELS OF WAR

2-11. SF units conduct operations in either an overt, low-visibility, or clandestine manner at the tactical level; however, the effects of these operations can have significance at the operational and strategic levels of war. All members of SF must pay close attention to how tactical tasks lead to campaign objectives and how those campaign objectives meet national security goals.

2-12. Operations at the strategic level focus on meeting the end-state objectives of national and theater policy. Decisions at this level reflect national and multinational goals, integrate the instruments of national power, provide forces, and determine constraints and restraints necessary for the effective use of available forces. The President, Secretary of Defense, and the GCCs determine the strategic national and theater objectives. They also determine the manner of military means to meet these objectives. The President, Secretary of Defense, or the GCCs may directly or indirectly (through subordinate commanders) integrate and employ SF in pursuit of these objectives.

2-13. The operational level focuses on theater campaigns and major operations. JFCs determine operational objectives that lead to the attainment of strategic theater objectives. These objectives are obtained through the design, organization, and conduct of campaigns and major operations, which in turn guide tactical events. A GCC or subordinate commanders (through the GCC) can request SF as part of the joint force organization to achieve these operational goals.

2-14. The tactical level of operations focuses on battles and engagements. Decisions at this level apply combat power to create advantages while in contact with or close to the enemy. SF may support tactical actions (offense, defense, and stability tasks) designed to have significant effects in obtaining operational objectives. Tactical-level operations normally rely on standard Army small-unit tactics but may include advanced methods of infiltration and/or exfiltration, as well as special equipment and techniques.

OPERATIONAL ENVIRONMENTS

2-15. SF conducts operations in all operational environments, which are defined as "a composite of the conditions, circumstances, and influences that affect the employment of capabilities and bear on the decisions of the commander" (JP 3-0). Permissive environments are those in which host-country military and law enforcement agencies have control, as well as the intent and capability, to assist operations that a unit intends to conduct. In uncertain environments, host-government forces, whether opposed or receptive to operations that a unit intends to conduct, do not have effective control of the territory and population in the intended operational area. These operational environments are not static. This ability of SF to work in both environments allows a tailored mix of the offensive, defensive, and stability tasks that results in deterrence of conflict or a favorable conflict resolution. Denied areas are areas that are operationally unsuitable for conventional forces because of political, tactical, environmental, or geographical reasons. It is a primary area for SOF (FM 3-05). Physical movement is what is commonly referred to when access or movement is denied, such as the East/West Wall that existed in Berlin. This is not the case when dealing in today's dynamic world with all of the variables that it has to offer, to include cyberspace.

Permissive Environment

2-16. SF provides a unique capability for the GCC's peacetime campaign strategy to gain or maintain U.S. access to strategically important foreign countries, to demonstrate U.S. commitment or resolve, or to otherwise contribute to collective security. An example of this includes sustained engagement through the theater security cooperation plan in which SF conducts joint combined exchange training, counternarcotics training, and other activities in support of the GCC. Strategies for peacetime SF operations are to prevent conflict through early intervention or to deter a crisis. SF units can be employed to help convince hostile powers to respect U.S. and HN national interests and to refrain from acts of international aggression and coercion.

Uncertain Environment

2-17. As crises develop, regional expertise and sustained engagement enable SF teams to assess, as well as shape, events. SF conducts a variety of missions based upon the GCC's campaign plan. Early use of SF to preempt or resolve a crisis can preclude the need to involve U.S. conventional forces. In addition, early use of SF can help set the conditions for successful rapid and decisive operations through preparation of the environment. The broad distribution of SF missions and training activities in a given theater during peacetime or periods of uncertainty make SF widely available to precede conventional forces into a hostile area.

2-18. In the early stages of developing crises, the commitment of conventional combat forces may be premature, inappropriate, or infeasible, and may risk further escalation. In these situations—when political, economic, and diplomatic means are inadequate to respond to a conflict—the SFGs give the President and

Secretary of Defense options for discriminate engagement precluding or limiting the need to employ conventional combat forces. The low visibility of SF operations helps the United States and its allies to maintain diplomatic flexibility. SF operations may also allow other powers (friendly, neutral, and hostile) to accept the outcome of multinational operations because they avoid the publicity of a more obvious use of military force.

2-19. When providing support to HNs and partner nations during FID, counterinsurgency, or counternarcotics operations, SF maintains a predominantly proactive (rather than reactionary) posture. Whenever suitable and feasible, SF organizes, trains, and supports (or directs) indigenous combat forces to locate and destroy hostile insurgent forces in contested areas. An example of this would be the partnership between SF and Afghan commando units. While some SF elements worked with Afghan local police in a primarily local defensive posture generating intelligence packets, SF and Afghan commando elements supported this by conducting limited offensive strikes against identified targets. Another example is the support SF has given to Colombian counterinsurgent and counternarcotics units, thereby enabling them to aggressively take the fight to the enemy.

Denied Areas

2-20. Whether already present or able to move quickly because of proximity, SF can usually be in a denied area well before conventional forces. SF can provide early critical information and ground truth to the commander and help set the conditions for conventional force operations unilateral special operations. Such operations may discreetly extend into the homeland of a hostile power. During war, the strategic role for SF is to focus on the long-term capacity of the hostile power to continue hostilities. SF may infiltrate into denied areas to support resistance forces and to collect and report information of national strategic importance.

2-21. The GCC may employ SF to interdict the advance of hostile forces to gain more time for employing conventional forces. SF may perform special reconnaissance tasks at the theater strategic level to identify hostile capabilities, intentions, and activities of importance to the GCC. As a strategic economy of force, SF can be employed to delay, disrupt, or harass hostile reinforcing forces or divert them to secondary areas of operations to alter the momentum and tempo of hostile operations. One of the core proficiencies of SF is the role of the combat advisor. In the context of UW, SF elements function as advisors and trainers to indigenous forces in denied areas.

SPECIAL FORCES OPERATIONS WITHIN THE RANGE OF MILITARY OPERATIONS

2-22. The United States employs SF capabilities in support of national security goals. These employments vary in size, purpose, and combat intensity within the range of military operations. These operations extend from military engagement, security cooperation, and deterrence, to crisis response and limited contingency operations and, if necessary, major operations and campaigns. In support of unified land operations—Army capstone doctrine—SF provides forces adept at running simultaneous offensive, defensive, and stability tasks or defense support of civil authorities. SF is further defined through the use of its critical capabilities: special warfare and surgical strike. *Special warfare* is the execution of activities that involve a combination of lethal and nonlethal actions taken by a specially trained and educated force that has a deep understanding of cultures and foreign language, proficiency in small-unit tactics, and the ability to build and fight alongside indigenous combat formations in a permissive, uncertain, or hostile environment. (ADP 3-05). *Surgical strike* is the execution of activities in a precise manner that employ special operations forces in hostile, denied, or politically sensitive environments to seize, destroy, capture, exploit, recover or damage designated targets, or influence threats (ADP 3-05). These capabilities are relevant and valuable throughout the entire range of military operations. Use of SF capabilities in military engagement, security cooperation, and deterrence activities helps shape the operational environment and can keep the day-to-day tensions between nations or groups below the threshold of armed conflict while maintaining U.S. global influence. Many of the missions associated with limited contingencies, such as foreign humanitarian assistance, do not require combat. Others, such as the humanitarian mission turned broader stability mission in Somalia in the early 1990s, for example, can be extremely dangerous and require a significant effort to protect U.S. forces while accomplishing the mission. Individual major operations and campaigns often contribute to a larger, long-term effort. For example, as significant as combat and other activities have been in Afghanistan and Iraq since 11 September 2001, they are only a part of global operations against terrorist networks. The nature of the security environment is such that SF is often engaged in several types of joint operations simultaneously.

ARMY UNIFIED LAND OPERATIONS

2-23. The term "unified land operations" describes the Army's warfighting doctrine. It is based on the central idea that the Army units seize, retain, and exploit the initiative to gain position of relative advantage over the enemy. This foundation of unified land operations is built on initiative, decisive action, and mission command. Army forces combine offensive, defensive, and stability tasks or defense support of civil authorities simultaneously to set the conditions for conflict resolution. This operational concept recognizes the three-dimensional nature of modern warfare and the need to conduct a fluid mix of offensive, defensive, and stability tasks. Unified effort among friendly forces—across all operational environments and against diverse and unpredictable threats—is critical to success.

2-24. The SF persistent presence around the world, the cultural expertise developed through education and sustained regional engagement, and advanced (often classified) skills make SF especially valuable during the shape and deter phases of an operational plan (Figure 2-1, page 2-7), and described in greater detail in ADP 3-05). As part of a dedicated UW campaign, SF, along with Army special operations forces (ARSOF), joint, and interagency partners, may even prevent situations from developing into major combat operations. Concurrently in these early phases, SF can conduct FID in countries adjacent to the targeted area, which can indirectly influence (shape) the situation.

2-25. Army special operations are executed throughout the range of military operations; however, Army special operations in the shape and deter phases focus on preventing conflict. The graphically extended shape phase depicted on both ends of the notional operational plan phases (Figure 2-1, page 2-7) highlights the disproportionate amount of time that should be spent in this phase. During the seize initiative and dominate phases, SF capabilities to unilaterally conduct direct action, special reconnaissance, counterterrorism, and other operations could be an integral part of the larger campaign plan. Initiative can be gained by organizing, leading, and amplifying the effects of indigenous partners and by integrating their activities with the larger campaign. UW or FID in support of larger conventional efforts provides the JFC with potentially crucial options in economy of force, surprise, maneuver, and otherwise unavailable intelligence advantages.

2-26. Developed relationships and operational experience between SF teams and their indigenous partners who supported U.S.-led major combat operations will ultimately make the JFC's task easier in the stabilize and enable civil authority phases. SF and other ARSOF in Civil Affairs (CA) and Military Information Support (MIS) are excellent resources for transitioning from conflict to a steady state. Although some combat roles for SF teams and their indigenous partners may continue in these phases—especially if the operation becomes a counterinsurgency effort—most likely SF will then concentrate on FID and assistance with integrating Civil Affairs operations (CAO) and Military Information Support Operations (MISO) to build capacity and confidence in the security sector, which will enable success in all other sectors needed for development. Sustained SF engagement in the complex postconflict environment provides the civil authorities with experience and mature forces that understand the population and can perceive the transitional threats before they become entrenched obstacles to stabilization.

SPECIAL FORCES–CONVENTIONAL FORCE COORDINATION AND INTEGRATION

2-27. Recent operations overseas often produced situations where conventional forces and SF operated simultaneously in the same area of operations and in close proximity. FM 6-05 establishes tactics, techniques, and procedures and aids conventional and SOF commanders and staffs. It provides the information and coordination required to ensure effective integration and interoperability between conventional forces and SOF.

The Role of Special Forces

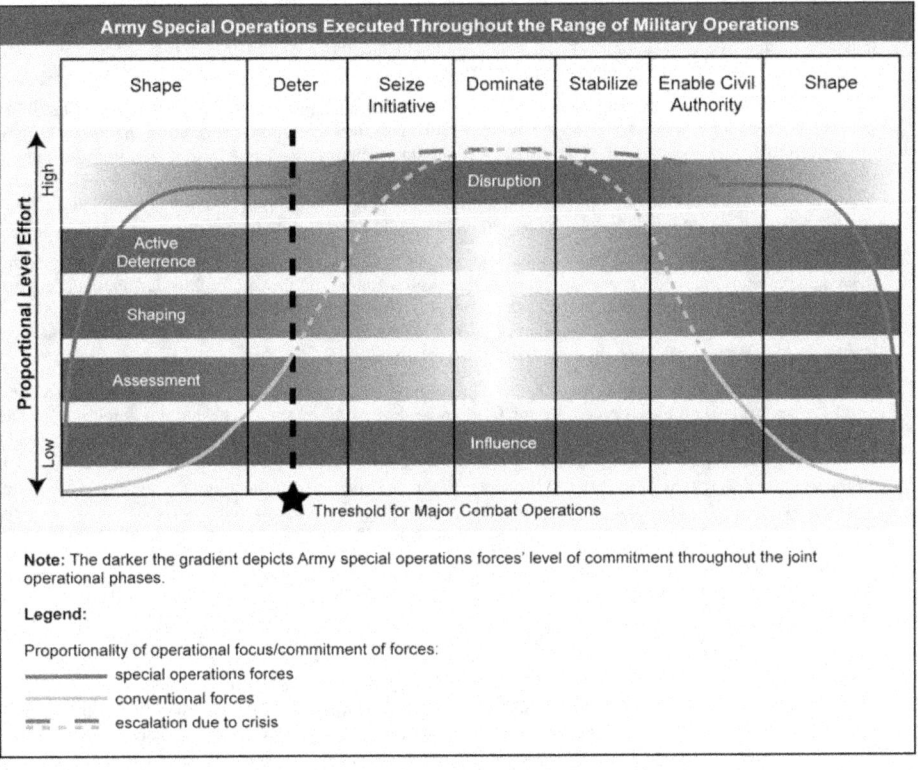

Figure 2-1. Role of Army special operations forces

2-28. In order to generate increased operational effectiveness, SF and conventional forces must not lose key lessons learned and must plan, coordinate, and synchronize operations to achieve the maximum effectiveness of their respective capabilities. This necessitates greater interoperability and, in specific situations, the integration of SF and conventional force processes, capabilities, and/or units to achieve operational objectives. A more cohesive SF and conventional forces effort to enhance interoperability and integration improves the ability to execute operations by combining the capability advantages of each force.

2-29. Key to the successful integration of SF and conventional forces is understanding and appreciating the capabilities and limitations of each force, especially by the leaders of each. This should include the capability to support differing missions, specific force limitations and shortfalls, HN and multinational force capabilities, and unique operational capabilities. Input from exchanged liaison officers is essential to provide this understanding to unit planners and staffs.

2-30. Commanders and staffs must understand the command and support relationships. Once JFC establishes the command and support relationships, the SF and conventional force commanders must clearly understand the supported and supporting command relationship. It is important that both commanders achieve a common understanding of the supported and supporting relationships during operations, fires, logistics, and unified action partners' integration.

2-31. Even when supporting and supported relationships are established, it is critical to clearly state what support is to be provided when and where by each force to the other, with particular emphasis on intelligence, fires, and sustainment support. Further, both forces must understand the control measures in place by each force for maneuver, fires, and aviation in order to ensure support is provided when and where needed and to avoid any possibility of fratricide or delayed support.

2-32. Conventional forces and SF units should meet and integrate early to foster the relationship, instill the *one team—one fight* mentality, understand each other's staff planning procedures, and defuse any misconceptions or friction points. If possible, units should attend training events together at training centers or as part of joint exercises. Best practices include traveling to one's home station, briefing each other's capabilities and limitations, mission goals, and linking up staff function sections. Ideally, a relationship is already established through a regional focus and habitual relationship with the corresponding TSOCs.

2-33. During operations, commanders must understand and synchronize each other's mission planning cycle, operations and intelligence processes, and mission approval processes. Mission type orders are optimal to convey the commander's intent to permit flexibility, initiative, and responsiveness. Coordinated rehearsals between conventional forces and SOF provide the best means of reducing missed opportunities, unnecessary delays, and the potential for fratricide during operations.

2-34. Conventional forces and SF intelligence sharing and collaboration must occur at every level to find and develop targets and aid in preparing for future operations. Information considered unimportant to one may be the critical missing piece for the other. Conventional forces and SF intelligence personnel should keep in constant contact with each other. To facilitate this constant sharing, organizations can establish fusion centers to integrate conventional forces and SOF information and operations. The information and intelligence produced and disseminated by a fusion center assists commanders and staffs in developing future operations (for example, targeting [nonlethal or lethal actions]).

2-35. Fusion centers mark a significant improvement to dynamic operational support by integrating mission command with focused analysis within a single centralized entity. Whereas intelligence fusion centers may be the most common or well known, these centers are not necessarily intelligence-led organizations. A fusion center is an ad hoc collaborative effort between several units, organizations, or agencies that provide resources, expertise, information, and intelligence to a center with the goal of supporting the rapid execution of operations by contributing members. These centers are primarily designed to focus collection and promote information sharing among multiple participants within a specific geographic area or mission type. These centers are not operations centers. Commanders at various echelons create fusion centers to manage the flow of information and intelligence, focus information collection to satisfy information requirements, and process, exploit, analyze, and disseminate the resulting collection.

THE NATURE AND LIMITATIONS OF SPECIAL FORCES

2-36. SF is the largest component of ARSOF. As such, the nature and limitations of SF are represented in a number of conceptual guidelines. These include operational characteristics and planning factor, special operations mission criteria, and ARSOF (SF) capabilities, characteristics, and imperatives. It is essential that JFCs commanding, cooperating with, or supporting SF clearly understand these criteria.

SPECIAL FORCES OPERATIONAL MISSION CRITERIA

2-37. The employment of SF in support of the joint force campaign or operation plan is facilitated by five basic operational mission criteria. During operational planning, SF planners must be integrated with the appropriate levels of authority to explain the capabilities and limitations of their units. The following five criteria provide guidelines for conventional and SF commanders and planners to use when considering the employment of SF:

- It must be an appropriate SF mission or activity. SF should be used to achieve effects that require its specific skills and capabilities. If the effects do not require those skills and capabilities, SF should not be assigned. SF should not be used as a substitute for other forces.
- The mission or tasks should support the JFC's campaign or operation plan or special activities. If the mission does not support the JFC's campaign or major operation plan, more appropriate missions available for SF should be considered instead.

- The mission or tasks must be operationally feasible. SF is not structured for attrition or force-on-force warfare and should not be assigned missions beyond their capabilities. SF commanders and their staffs must consider the vulnerability of SF units to larger, more heavily armed or mobile forces, particularly in hostile territory.
- Required resources must be available to execute the mission. Some SF missions require support from other forces for success. Support involves aiding, protecting, complementing, and sustaining employed SF. Support can include airlift, intelligence, communications, fires, information operations, medical, logistics, space, weather, and security. Although a mission may be appropriate for SF, insufficient resources may invalidate employing SF for that mission.
- The expected outcome of the mission must justify the risks. Commanders must make sure the benefits of successful mission execution justify the inherent risks. Risk management considers not only the potential loss of SF units and equipment but also the risk of adverse effects to U.S. diplomacy because of mission failure. Although SF may offer powerful tools for the JFC's campaign or operation, there may be some operations that SF can and will execute that make only a marginal contribution.

SPECIAL FORCES CAPABILITIES

2-38. The signature capabilities of SF are a function of the quality of SF Soldiers, the training and education of those Soldiers, and the mission profiles the Soldiers must execute. The challenging SF selection process, coupled with advanced tactical training, language skills, and education produce an SF Soldier who is adaptable, mature, innovative, culturally aware, self-assured, and self-reliant. Thus, policy decisionmakers use SF to expand the range of available options. SF skills enable it to work as effectively with civilian populations as with other military forces to influence situations favorably toward U.S. national interests. This ability to apply discreet leverage is an invaluable SF contribution to the National Security Strategy.

2-39. SF is a specially organized, trained, and equipped military force. It conducts special operations to achieve military, political, economic, or informational objectives by generally unconventional means in hostile, denied, or politically sensitive areas. SF operations differ from conventional force operations by their degree of acceptable physical and political risk, their modes of employment, and their operational techniques. SF may allow the GCC or JFC to achieve objectives typically beyond the capabilities of conventional forces.

2-40. SF expands the options of the President, Secretary of Defense, country teams, and GCCs, particularly in crises and contingencies that fall between wholly diplomatic initiatives and the overt use of large conventional forces. The small size, rapid-reaction capability, and self-sufficient nature of SF elements provide military options that do not involve the risk of escalation normally associated with larger, more visible conventional forces. The use of SF provides decisionmakers with an option to prevent a conflict or to limit its scope.

SPECIAL FORCES CHARACTERISTICS

2-41. To ensure missions selected for SF are compatible with their capabilities, commanders must be familiar with the following SF characteristics:
- SF personnel undergo careful selection processes or mission-specific training beyond basic military skills to achieve entry-level special operations skills. Being proficient in these skills makes rapid replacement or generation of personnel or capabilities highly unlikely.
- Mature, experienced personnel make up SF. Many SF Soldiers maintain a high level of competency in more than one military specialty.
- Most SF Soldiers are regionally oriented for employment. Cross-cultural communication skills are a routine part of their training.
- SF conducts specific tactical operations by small units with unique talents that directly strike or engage strategic and operational aims or objectives.

- Planning for SF operations may begin at the combatant command, joint, or interagency level for execution that requires extensive, rigorous rehearsal.
- SF operations are frequently clandestine or low-visibility operations, or they may be combined with overt operations. SF operations can be covert but require a declaration of war or a specific finding approved by the President or the Secretary of Defense. SF can deploy at relatively low cost with a low profile less intrusive than that of larger conventional forces.
- Selected SF units often conduct activities at great distances from operational bases. These units employ a sophisticated communications system and means of insertion, support, and extraction to penetrate and return from hostile, denied, or politically sensitive areas.
- SF operations occur throughout the range of military operations.
- SF operations influence the will of foreign leadership to create conditions favorable to U.S. strategic aims and objectives.
- SF operations are often high-risk operations that have limited windows of execution and require first-time success.
- Employment of SF may require patient, long-term commitment and support to achieve U.S. national goals in an area of operations. SF is ideally suited to perform operations through or with indigenous populations.
- SF operations require theater and, frequently, national-level intelligence support.
- Selected SF operations require a detailed knowledge of the cultural nuances and languages of a country or region where employed.
- SF operations are inherently joint and sometimes multinational, requiring interagency and international coordination. The contribution of SF to national security is greatest when SF is fully integrated into the JFC's plan at the earliest stages of planning.
- SF can be task organized quickly and deployed rapidly to provide scalable responses to situations.
- SF can gain access to hostile and denied areas.
- SF Soldiers can provide limited security and medical support for themselves and commonly rely on indigenous capabilities.
- SF can live in austere, harsh environments without extensive support. For long-duration operations, SF requires support from the Army Service component command.
- SF can survey and assess local situations and rapidly report these assessments.
- SF works closely with regional military and civilian authorities and populations.

ARMY SPECIAL OPERATIONS FORCES IMPERATIVES

2-42. The following 12 ARSOF imperatives from ADRP 3-05 also apply to SF. These imperatives are the foundation for planning and executing special operations in concert and integrated with other forces, interagency partners, and foreign organizations. SF includes the applicable imperatives in their mission planning and execution. The imperatives are—

- **Understand the Operational Environment.** SF cannot shape the operational environment without first gaining a clear understanding of the joint operations area, including civilian influence and enemy and friendly capabilities. SF applies the political, military, economic, social, information, infrastructure, physical environment, time operational variables to analyze the operational area. SF must identify the friendly and hostile decisionmakers, their objectives and strategies, and the ways they interact. The conditions of conflict can change and SF must anticipate these changes in the operational environment and exploit fleeting opportunities.
- **Recognize Political Implications.** Many SF operations are conducted to advance critical political objectives. SF must understand that its actions can have international consequences. Rules of engagement provide a framework that considers political implications. However, rules of engagement cannot anticipate every situation. SF must understand the intent of the rules of engagement and act accordingly, despite any military disadvantage that may result. The advancement of the political objective may take precedence over the military disadvantages.

- **Facilitate Interagency Activities.** SF must actively and continuously coordinate its activities with all relevant parties (U.S. and foreign military and nonmilitary organizations) to ensure efficient use of all available resources and maintain unity of effort.
- **Engage the Threat Discriminately.** SF discerns differences in the threats and engages each accordingly. A direct action solution for one threat may be inappropriate for another, especially if that threat could be bypassed or marginalized without loss or co-opted for great material or political advantage.
- **Consider Long-Term Effects.** SF must consider the political, economic, informational, and military effects when faced with dilemmas because the solutions will have broad, far-reaching effects. SF must accept legal and political constraints to avoid strategic failure while achieving tactical success. SF must not jeopardize the success of national and GCC long-term objectives by the desire for immediate or short-term effects. SF policies, plans, and operations must be consistent with the national and theater priorities and objectives they support. Inconsistency can lead to a loss of legitimacy and credibility at the national level.
- **Ensure Legitimacy and Credibility of Special Operations.** Significant legal and policy considerations apply to many SF activities. Legitimacy is the most crucial factor in developing and maintaining internal and international support. The United States cannot sustain its assistance to a foreign power without this legitimacy. The concept of legitimacy is broader than the strict legal definition contained in international law. The concept also includes the moral and political legitimacy of a government or resistance organization. The people of the nation and the international community determine its legitimacy based on collective perception of the credibility of its cause and methods. SF, as a frequent choice as an early-entry force, must ensure its actions and communications with the local populace and forces it is working with are consistent with stated U.S. policy. Consistency of message and action is critical to legitimacy and credibility. Without legitimacy and credibility, SF will not gain the support of foreign indigenous elements, the U.S. population, or the international community. SF legal advisors must review all sensitive aspects of SF mission planning and execution.
- **Anticipate and Control Psychological Effects.** All special operations have significant psychological effects that often are amplified by an increasingly pervasive electronic media environment and the growing influence of social media. SF must integrate MISO and public affairs into its activities, anticipating and countering threat information themes, as well as managing the second- and third-order effects of operations on a population.
- **Operate With and Through Others.** The primary role of SF in multinational operations is to advise, train, and assist indigenous military and paramilitary forces. The supported non-U.S. forces then serve as force multipliers in the pursuit of mutual security objectives with minimum U.S. visibility, risk, and cost. SF also operates with and through indigenous government and civil society leaders to shape the operational environment. The long-term self-sufficiency of the HN militaries and civil authorities must assume primary authority and accept responsibility for the success or failure of the mission. All U.S. efforts must reinforce and enhance the effectiveness, legitimacy, and credibility of the supported foreign government or group.
- **Develop Multiple Options.** ARSOF maintain their agility by developing a broad range of options and contingency plans to provide flexible national and regional options. Keys to operational flexibility include—
  - Developing contingency plans that anticipate problems during critical events.
  - Using a collaborative, deliberate, and interactive planning and rehearsal process.
  - Having the same people plan, rehearse, and execute the mission. These types of actions on the objective become a common point of departure, not inflexible blueprints.

Note: Under these circumstances, the participants understand all the critical elements of the plan, as well as alternate courses of action, reasons for discarding alternate courses of action, and unstated assumptions underlying unexpected difficulties.

- **Ensure Long-Term Engagement.** SF recognizes the need for persistence, patience, and continuity of effort in addressing security issues abroad. When supporting a U.S. policy of enduring engagement, planners and those executing missions must work to ensure programs are sustainable (within the capabilities of the HN), consistent (properly timed to maintain continuity), and durable (sufficiently resourced).
- **Provide Sufficient Intelligence.** Success for SF missions dictates that uncertainty associated with the threat and environment must be minimized through the application of intelligence operations and procedures. Because of the requirement for detailed intelligence, SF typically accesses theater and national systems to ensure that complete and predictive intelligence is applied. SF units also provide intelligence through area assessments, special reconnaissance, and post-operational debriefing of units. Human intelligence is often the only source that can satisfy critical intelligence requirements.
- **Balance Security and Synchronization.** Security concerns often dominate special operations, but overcompartmentalization can exclude key special operations forces, conventional forces, and indigenous personnel from the planning cycle. Insufficient security may compromise a mission. Excessive security may cause the mission to fail because of inadequate coordination. SF must constantly balance the two and resolve these conflicting demands on mission planning and execution.

Chapter 3
Guidance and Principal Tasks

Unconventional warfare . . . remains uniquely Special Forces'. It is the soul of Special Forces: the willingness to accept its isolation and hardships defines the Special Forces Soldier. Its training is both the keystone and standard of Special Forces training: it has long been an article of faith, confirmed in over forty years of worldwide operations, that if you can do the unconventional warfare missions, you can do all others. The objective of unconventional warfare and Special Forces' dedication to it is expressed in Special Forces' motto: De Oppresso Liber (to free the oppressed).

Robert M. Gates
Remarks at the dedication of the Office of Strategic Services Memorial, Langley, VA
12 June 1992

This chapter discusses the generation of the SF nine principal tasks from the ARSOF 11 core activities and provides an overview of current doctrine for each. Although discussed separately, the nine principal tasks are all interrelated and reflect the inherent modifications and additions to training, organization, and equipment needed to meet the operational requirements of the GCCs.

The U.S. Army Special Operations Command (USASOC) organizes, trains, and equips SF to perform its principal tasks of UW, FID, security force assistance, counterinsurgency, preparation of the environment, direct action, special reconnaissance, counterterrorism, and counterproliferation of weapons of mass destruction. Through the conduct of these principal tasks, SF operations support the accomplishment of USASOC's specified SOF core activities. SF missions are dynamic and constantly evolving in response to political-military considerations, technology, and other considerations. A change in national security policy, National Military Strategy, global or regional social structure, or technology may radically alter the manner in which SF conducts its principal tasks.

GUIDANCE TO CONDUCT SPECIAL FORCES OPERATIONS

3-1. Special operations activities are founded in law and guided by joint and Army doctrine. Command structures ensure special operations units are trained and equipped to conduct their respective activities and tasks. The following paragraphs describe the higher military doctrine, command structures, and authorities from which SF units derive their nine principal tasks.

JOINT PUBLICATION 3-05

3-2. JP 3-05 provides fundamental principles and guidance for the Services, combatant commanders, and subordinate JFCs to prepare for and conduct special operations. It describes those military operations and provides general guidance for military commanders to employ and execute command and control of SOF assigned or attached to a GCC, subordinate unified commander, or a JTF commander.

ARMY DOCTRINE PUBLICATION 3-0

3-3. ADP 3-0 describes how the Army seizes, retains, and exploits the initiative to gain and maintain a position of relative advantage in sustained land operations through simultaneous offensive, defensive, and stability tasks in order to prevent or deter conflict, prevail in war, and create the conditions for favorable conflict resolution. It provides fundamental guidance for the three-dimensional nature of modern warfare and the need to conduct a fluid mix of offensive, defensive, and stability or defense support of civil authorities tasks simultaneously. It acknowledges that strategic success requires the full integration of U.S. military operations with interagency and multinational partners. The doctrine supports the use of SF as an adaptable component of land operations that can conduct simultaneous operations with multiple partner forces and agencies in support of comprehensive campaigns involving offense, defense, stability tasks, or defense support of civil authorities.

3-4. ADP 3-05 describes the role of U.S. ARSOF in the U.S. Army's operating concept to shape operational environments in the countries and regions of consequence, prevent conflict through the application of special operations and conventional deterrence, and, when necessary, help win our nation's wars. It outlines the ARSOF requirement to provide in the nation's defense unequalled surgical strike and special warfare capabilities. Together these two different but mutually supporting forms of special operations comprise the American way of special operations warfighting.

3-5. ADRP 3-05 provides a broad understanding of Army special operations by describing how executing the two mutually supporting critical capabilities of special warfare and surgical strike contribute to unified land operations. ADRP 3-05 provides a foundation for how the Army meets the JFC's needs by appropriate integration of ARSOF and conventional forces.

FIELD MANUAL 3-05

3-6. FM 3-05 describes the ARSOF strategic landscape, fundamentals, core activities, capabilities, and sustainment involved in the range of military operations and serves as the doctrinal foundation for subordinate ARSOF doctrine, force integration, materiel acquisition, professional education, and individual and unit training. FM 3-05 provides the linkage from joint special operations and Army doctrine to ARSOF doctrine and provides guidance for ARSOF commanders who determine the force structure, budget, training, materiel, and operational requirements necessary to prepare ARSOF to conduct their core activities.

UNITED STATES SPECIAL OPERATIONS COMMAND

3-7. Created by the Goldwater-Nichols Department of Defense Reorganization Act of 1986 and the Nunn-Cohen Amendment to the National Defense Authorization Act of 1987, USSOCOM was activated on 16 April 1987 at MacDill Air Force Base, Florida, in order to prepare SOF to carry out assigned missions and, if so directed, to plan for and conduct special operations. *Special operations* are operations requiring unique modes of employment, tactical techniques, equipment and training often conducted in hostile, denied, or politically sensitive environments and characterized by one or more of the following: time sensitive, clandestine, low visibility, conducted with and/or through indigenous forces, requiring regional expertise, and/or a high degree of risk (JP 3-05).

3-8. SOF are those forces identified in Title 10, U.S. Code, Section 167, *Unified Combatant Command for Special Operations Forces*, or those units or forces that have since been designated as SOF by the Secretary of Defense. Under Title 10, the Commander, USSOCOM, shall be responsible for and shall have the authority to conduct the functions relating to special operations activities. These functions include training, equipping, and providing SOF for the GCCs.

3-9. USSOCOM Directive 10-1cc, *Terms of Reference—Roles, Missions, and Functions of Component Commands*, provides terms of reference to the component commands in order to identify command relationships, missions, functions, and responsibilities. Headquarters, USSOCOM, through its component and subunified commands, specifically organizes, trains, and equips SOF to accomplish the following 11 core activities:

- UW.
- FID.
- Security force assistance.
- Counterinsurgency.
- Direct action.

- Special reconnaissance.
- Counterterrorism.
- MISO.
- CAO.
- Counterproliferation of weapons of mass destruction.
- Information operations.

UNITED STATES ARMY SPECIAL OPERATIONS COMMAND

3-10. USASOC possesses the capabilities to support USSOCOM's roles, missions, and functions as directed by Congress in Title 10, U.S. Code, Section 164, *Commanders of Combatant Commands: Assignment; Powers and Duties;* and Title 10, U.S. Code, Section 167, *Unified Combatant Command for Special Operations Forces.*

3-11. USASOC provides trained and ready SF, Ranger, special operations aviation, MIS, and CA personnel to GCCs and U.S. Ambassadors. Commander, USASOC, exercises command of continental United States-based Regular Army SOF. He also oversees and evaluates continental United States-based Army National Guard SOF. USASOC is responsible for the development of ARSOF doctrine; tactics, techniques, and procedures; and materiel. USASOC is organized, trained, and equipped specifically to build the capability to accomplish the following 12 core activities:

- UW.
- FID.
- Security force assistance.
- Counterinsurgency.
- Preparation of the environment.
- Direct action.
- Special reconnaissance.
- Counterterrorism.
- Counterproliferation of weapons of mass destruction.
- MISO.
- CAO.
- Humanitarian assistance/disaster relief.

UNITED STATES ARMY SPECIAL FORCES COMMAND

3-12. The U.S. Army Special Forces Command (USASFC) comprises five Regular Army SFGs and two Army National Guard SFGs. When tasked by USASOC to fill personnel requirements for validated missions in current and projected war plans, USASFC may recommend both Regular Army and Army National Guard units and direct training priorities to meet requirements for assigned missions. USASFC and its SFGs transform this guidance into mission-essential tasks for the battalion, company, and detachment based on collaboration with subordinate units and mission analysis.

SPECIAL FORCES PRINCIPAL TASKS

3-13. SF is organized, manned, trained, and equipped to execute its nine principal tasks. These tasks are derived from the 12 core activities of USASOC. Although conventional forces also conduct many of these tasks (for example, FID, security force assistance, counterinsurgency, and counterterrorism), SF conducts these tasks using specialized formations, tactics, techniques, and procedures. Use of SF with conventional forces creates an additional and exceptional capability to achieve objectives that may not be otherwise attainable. SF can arrange and package its capabilities in combinations to provide DOD options applicable to a broad range of strategic and operational challenges. Additionally, using the capabilities developed to conduct the principal tasks, SF can perform other tasks of a collateral nature, such as counterdrug operations and noncombatant evacuation operations. Preparation of the environment is a specialized core

Chapter 3

task used as a shaping activity that supports all special operations. Figure 3-1 depicts the SF principal tasks arrayed against the USASOC two critical capabilities—special warfare and surgical strike. Special warfare capabilities emphasize working with HNs, regional partners, and indigenous populations. Those tasks generally include UW, FID, counterinsurgency, and security force assistance. On the other hand, surgical strike is executed by certain small-unit operations with extensive training for extreme risk and precise execution while delivering high-payoff results. Within SF, only the crisis force is trained to execute surgical strike missions. Those missions are generally counterterrorism, counterproliferation, unilateral direct action, and special reconnaissance. Preparation of the environment supports both capabilities.

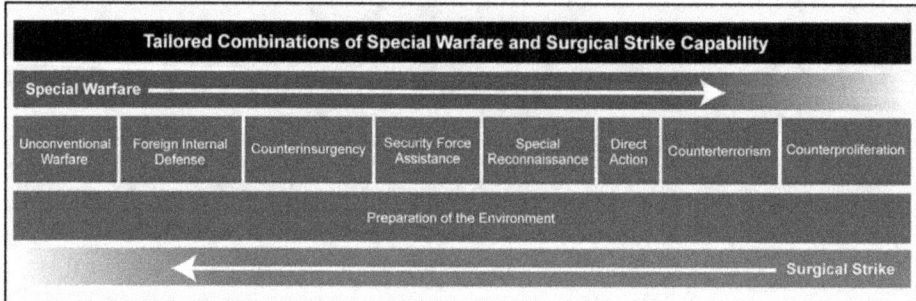

Figure 3-1. Special Forces principal tasks

UNCONVENTIONAL WARFARE

3-14. UW is the core task and organizing principle for Army SF. UW capabilities provide the method and skill sets by which all other SF missions are accomplished. SF is specifically organized, trained, and equipped for the conduct of UW. SF is regionally oriented, language-qualified, and specifically trained to conduct UW against hostile nation-states and non-state entities to achieve U.S. goals.

3-15. UW is defined as "activities conducted to enable a resistance movement or insurgency to coerce, disrupt, or overthrow a government or occupying power by operating through or with an underground, auxiliary, and guerrilla force in a denied area." The United States may engage in UW as part of a major theater war or limited regional contingency as an effort to support an insurgency or resistance movement. Experiences in the 1980s in Afghanistan and Nicaragua proved that support for an insurgency could be an effective way of putting indirect pressure on the enemy. The cost versus benefit of using UW must be carefully considered before employment. Properly integrated and synchronized UW operations can extend the application of military power for strategic goals. UW complements operations by giving the United States opportunities to seize the initiative through preemptive or clandestine offensive action.

3-16. UW is inherently a joint and interagency activity. For its part, SF units are designed to accomplish the following significant aspects:
- Infiltrating denied territory and linking up with resistance forces.
- Assessing resistance forces for potential sponsorship by the U.S. government.
- Providing training and advisory assistance to the guerrilla forces or the underground.
- Coordinating and synchronizing the resistance activities with U.S. efforts.
- Transitioning guerrilla forces into post-conflict status as well as anticipating and influencing situations where former resistance elements could develop into an armed insurgency against the newly formed U.S.-sponsored government.

3-17. Planning for UW is different from planning for other special operations. UW involves long-term campaigns that require operational art to put forces in space and time and integrate ends, ways, and means that attain the desired U.S. political or military end states. The sensitivity of the planned action dictates the level of compartmentalization the United States must use to ensure operational security.

Parallel planning by all levels ensures that each level understands how their mission integrates with the missions of other levels.

3-18. Military leaders must carefully consider the costs and benefits prior to making a decision to employ UW. Properly integrated and synchronized UW complements other operations by giving the United States or HN opportunities to seize the initiative through preemptive covert or clandestine offensive action without an overt commitment of a large number of conventional forces.

3-19. The goal of UW operations is a change in political control and/or perceived legitimacy of regimes. Hence, UW has strategic utility that can alter the balance of power between sovereign states. Such high stakes carry significant political risk in both the international and domestic political arenas and necessarily require sensitive execution and oversight. The necessity to operate with a varying mix of clandestine and covert means, ways, and ends places a premium on excellent intelligence of the UW operating area. In UW, as in all conflict scenarios, SF must closely coordinate activities with joint, interagency, intergovernmental, and multinational partners in order to enable and safeguard sensitive operations.

3-20. A TSOC typically tasks SF to lead a UW campaign. UW will usually require some interagency support and possibly some support by conventional forces. The prevailing strategic environment suggests a TSOC and staff must be able to effectively conduct and support UW simultaneously during both traditional warfare and irregular warfare. In some cases, SF conducting UW will be the main effort, with conventional forces playing a much smaller and supporting role.

3-21. Each instance of UW is unique; however, UW efforts generally pass through seven distinct phases: preparation, initial contact, infiltration, organization, buildup, employment, and transition. These phases may occur simultaneously in some situations or may not occur at all in others. For example, a large and effective resistance movement may only require logistical support, thereby bypassing the organization phase. The phases may also occur out of sequence, with each receiving varying degrees of emphasis. One example of this is when members of an indigenous irregular force are moved to another country to be trained, organized, and equipped before being infiltrated back into the UW operations area.

Phase I: Preparation

3-22. The preparation phase for UW is part of a three-step process that consists of intelligence preparation of the battlefield, war planning, and shaping activities. First, intelligence preparation of the battlefield attempts to graphically represent the current reality and predict probable enemy courses of action in the UW operating area. Second, war planning describes possible or probable future military operations. Third, shaping activities work to modify the UW operating area to make it more conducive to all types of future operations. SF personnel should be cautious and deliberate using terms regarding preparation to avoid confusion or misinterpretation. Intelligence preparation of the battlefield is an Army term; it is not a joint term and certainly is not an interagency term. Intelligence preparation of the operational environment is the joint term. Preparation of the environment is not synonymous with either.

Phase II: Initial Contact

3-23. Ideally, a pilot team should make initial contact with an established or potential irregular element. A pilot team is a deliberately structured composite organization composed of SF operational detachment members, with likely augmentation by interagency or other special skilled personnel, designed to infiltrate a designated area to conduct sensitive preparation of the environment activities and assess protraction of a resistance movement or insurgency for the conduct of UW in support of U.S. objectives. This team might also include members from multinational or partner nations. During initial contact, pilot team personnel begin to assess the potential of conducting UW in the UW operating area and the compatibility of U.S. and indigenous irregular element interests and objectives with the UW operating area. This procedure allows for an accurate assessment of UW feasibility in the UW operating area and arranges for the reception and initial assistance of additional elements, typically SF. If deemed necessary, the TSOC may arrange to exfiltrate an asset from the UW operating area to brief the U.S. planning elements during mission planning.

Phase III: Infiltration

3-24. During this phase, SF Soldiers infiltrate the UW operating area. Infiltration may be as overt as using a chartered civilian flight or as discreet as a clandestine insertion. The mission requirement—analyzed with respect to mission, enemy, terrain and weather, troops and support available, time available, and civil considerations—determines the most desirable method of infiltration. The infiltration phase concludes when follow-on SF units link up with the pilot team or irregular force. During and following infiltration, SF Soldiers continue the area assessment initiated by the pilot team to confirm or refute the information previously reported. Throughout the mission, SF must continue to report all relevant operational information to higher headquarters.

Phase IV: Organization

3-25. During the organization phase, SF Soldiers determine whether the organization of the resistance or insurgency is appropriate for the conditions of the UW operating area. As necessary, SF then reorganizes the resistance or insurgency elements and begins to develop the capability of the irregular force. Depending on the size and scope of the effort, the size of this force can range from one individual to a resistance element of potentially any size. Planners traditionally conceive UW with an emphasis on guerrilla warfare. Such efforts may entail the organization of guerrilla units to conduct combat operations. However, UW is not solely a guerrilla warfare effort. The organization of resistance or insurgent elements may involve other SOF personnel, such as MIS, CA, or other skill sets. Although each irregular force or organization in UW is unique, the traditional SF practice of conceiving UW as U.S. support to an insurgency or resistance movement provides a unifying general concept for irregular force organizational structure. SF attempts to subdivide an insurgency into three components—guerrillas, underground, and auxiliary. These components are designed to work together toward the same end.

3-26. The area command is not a separate physical node like the three standard components of the insurgency. The area command is instead integrated throughout these components at all levels of the irregular organization. The area command is made up from the leadership cells of the underground, auxiliary, and guerrillas, as well as members of SF when present in the UW operating area. Many of these leaders may function as deliberate or de facto leaders of a shadow government within the UW operating area or as a government-in-exile, if it exists. Key movement leaders who provide strategic political direction may also be identifiable in the shadow government or government-in-exile.

3-27. The guerrillas represent the most commonly recognized portion of the insurgency. They are a group of irregular, predominantly indigenous personnel organized along military lines to conduct paramilitary operations in enemy-held, hostile, or denied territory. Guerrillas carry out most of the armed conflict that openly challenges the regional authority.

3-28. The underground is a cellular organization within the irregular movement that is responsible for subversion, sabotage, intelligence collection, and other compartmentalized activities. Most underground operations will take place in and around population centers. As such, the underground must have the ability to conduct operations in areas that are usually inaccessible to the guerrillas, such as areas under government military control. Underground members often fill leadership positions, overseeing specific functions that auxiliary workers carry out. The underground and auxiliary—although technically separate units—are, in reality, loosely interconnected elements that provide coordinated capabilities for the irregular movement. The key distinction between them is that the underground is the element of the irregular organization that operates in areas denied to the guerrilla force.

3-29. The auxiliary is the primary support element of the irregular organization. The structure and operations of the auxiliary are clandestine in nature and its members do not openly indicate their sympathy or involvement with the irregular movement. This support enables the guerrilla force, and often the underground, to survive and function. This support can take the form of logistics, labor, or intelligence. Members of the auxiliary are sometimes characterized as "part-time members" of the irregular organization, continuing to participate in the life of their community. The populace appears concerned only with their normal occupations while at the same time engaging in irregular operations to varying degrees.

3-30. The mass base is the larger relevant indigenous population from which the irregular forces are drawn. Although not traditionally considered a component of a U.S.-sponsored insurgent or resistance movement, the mass base must be an organizational consideration. UW is an elemental irregular warfare activity; as such, influence over a relevant indigenous population is critical. Communist insurgents of the 20th century, such as those under the aegis of Chinese leader Mao Tse-tung, referred to the general population (in a UW operating area) as the "mass base." Insurgent leaders tasked elements of the underground with infiltrating civil institutions and manipulating popular grievances and overt indigenous political activities to support insurgent objectives. With the appropriate levels of approval and planning integration, the core SOF task of MISO could be utilized to develop and execute programs to increase the moral and physical support of the relevant mass base audiences for the irregular forces while simultaneously countering the propaganda and information activities of the threat.

Phase V: Buildup

3-31. The buildup phase involves the expansion of the irregular elements and their capabilities to meet mission objectives. SF tasks include infiltration or procurement of equipment and supplies to support this expansion and subsequent operations. During the buildup phase, SF assists the indigenous cadre in expanding into an effective organization that is capable of conducting operations against the existing government or occupying forces.

Phase VI: Employment

3-32. During the employment phase, indigenous or other irregular forces increasingly operate in a combat environment. These operations build in scope and size to support the joint objectives of the U.S. Government and the resistance movement or insurgency to overthrow the government or occupying force. If UW operations are part of a larger theater operation, SF ensures that the effects of these operations continue to support the goals of the theater commander or the GCC. These operations range from interdiction with guerrilla forces designed to drain the morale and resources of the hostile power through combat to active intelligence collection with an indigenous informant network. Regardless of the type of operation, the overall purpose is to achieve strategic political-military objectives.

Phase VII: Transition

3-33. Transition is the final, most difficult, and most sensitive phase of UW operations. The planning for transition begins when the U.S. Government decides to sponsor an irregular organization and ends in the UW operating area upon cessation of operations. Transition does not necessarily mean demobilization or the commencement of FID operations. Successful transition from combat operations requires a comprehensive approach to stability tasks where the drawdown of irregular indigenous forces support lines of efforts in security, justice and reconciliation, humanitarian assistance, governance and participation, and economic stabilization and infrastructure. Achieving the desirable end state (stable governance, sustainable economy, and safe and secure environment) will involve diverse objectives. Those distinctive SF characteristics, such as cross-cultural communication skills and interagency/international coordination, will prove highly useful in ensuring the transition from relatively simple warfighter objectives to complex stability objectives that support civil authority.

FOREIGN INTERNAL DEFENSE

3-34. JP 3-22 defines FID as "participation by civilian and military agencies of a government in any of the action programs taken by another government or other designated organization to free and protect its society from subversion, lawlessness, insurgency, terrorism, and other threats to its security." Like UW, FID is an umbrella concept that covers a broad range of activities. Its primary intent is to help the legitimate host government address internal threats and their underlying causes. Commensurate with U.S. policy goals, the focus of all U.S. FID efforts is to support the HN program of internal defense and development.

Chapter 3

Role of Special Forces in Foreign Internal Defense

3-35. U.S. military involvement in FID has traditionally focused on support of HN counterinsurgency efforts of allies and friendly nations. Counterinsurgency remains an important aspect of military FID operations. However, the primary SF mission in FID is to organize, train, advise, assist, and improve the tactical and technical proficiency of the HN forces. The major difference in the way that SF and conventional forces conduct FID operations is in the area of advisory operations. Although conventional forces conduct a great deal of training in support of HN forces, their role as advisors is like an additional duty. Their core training is in a specific field, whereas SF Soldiers are trained and expected to be advisors to a resistance.

3-36. As a force multiplier, SF units have and maintain advanced skills and capabilities (such as language) that enable them to conduct advisory operations with the HN for extended periods. Improved proficiency enables the HN forces to defeat internal threats to their stability, thereby limiting direct U.S. involvement. The emphasis is on training HN cadres, who will in turn train their compatriots. The capabilities that SF employs to perform its FID mission are those inherent to its UW mission. Only the operational environment is changed. The USSOCOM is the only combatant command with a legislatively mandated FID mission.

3-37. An SF FID mission may require assets ranging from a single SF team to a reinforced SFG. In the early stages of a nation's need for assistance, the level of SF participation may be as small as one ODA. In the more advanced stages, an SF company or battalion may establish an operational base (within or outside of country) and exercise operational control of SF units. Operational and support elements may be assigned to the base on a rotational or permanent basis. When the entire SFG deploys to the country, it normally establishes a special operations task force (SOTF). The SOTF may then elect to establish one or more SF advanced operations bases (AOBs). SF units participate in a variety of operations to accomplish their FID mission. The HN needs as well as the U.S. and HN agreements will dictate the quantity and level of support required to support the HN internal defense and development program. Internal defense and development is the full range of measures taken by a nation to promote its growth and to protect itself from subversion, lawlessness, insurgency, terrorism, and other threats to its security. Figure 3-2 depicts the role of SF in both UW and FID.

Unconventional Warfare	Foreign Internal Defense
Coerce, disrupt, or overthrow a government or occupying power: • Policy option. • Through or with indigenous force. • Subversion/sabotage.	Improve a nation-state's security apparatus: • Train, advise, and assist. • Primarily counterinsurgency-focused.
Degrade legitimacy and destabilize.	Reinforce legitimacy and stabilize.
Note: Enabled by Army special operations forces, conventional forces, and joint, interagency, intergovernmental, and multinational capabilities.	*Note:* Collaborative with Army special operations forces, conventional forces, and joint, interagency, intergovernmental, and multinational capabilities.

Figure 3-2. Role of Special Forces in unconventional warfare and foreign internal defense

3-38. FID is not restricted to times of conflict. It can also take place in the form of training exercises and other activities that show U.S. resolve to and for the region. These exercises train the HN to deal with potential internal threats. FID usually consists of indirect assistance, such as participation in combined exercises and training programs, or limited direct assistance without U.S. participation in combat operations.

Joint and Multinational Exercises

3-39. Exercises conducted are designed to support the GCC's objectives within a specific theater or region. They are conducted to improve relations, enforce U.S. commitment to the region, improve interoperability with HN forces, and enhance U.S. warfighting skills. These exercises can be Chairman of the Joint Chiefs of Staff-, GCC-, and Service-sponsored events.

3-40. There are various programs and exercises SF personnel can be involved with when supporting military FID operations. These missions can involve training, advising, and involvement in exercises sponsored through Department of State and DOD initiatives.

3-41. A program specific to SF is the joint combined exchange training program. The program is designed to train the SOF of the combatant command and is authorized under Title 10, U.S. Code, Section 2011, *Special Operations Forces: Training With Friendly Foreign Forces*. SF teams gain most of their experience training HN forces through the joint combined exchange training program. Added benefits to SF are regional familiarity, cross-cultural understanding, and access to numerous countries throughout the world not normally afforded to conventional forces.

COUNTERINSURGENCY

3-42. Described in FM 3-24 and JP 3-24, counterinsurgency is defined as "comprehensive civilian and military efforts taken to defeat an insurgency and to address any core grievances." SF contributes to counterinsurgency by providing highly adaptable and regionally experienced teams with the ability to operate discreetly in local communities through indigenous security forces. SF conducts counterinsurgency utilizing a balanced approach, simultaneously incorporating counterguerrilla, information collection, counterintelligence, and other operations as necessary to secure the general population and foster legitimacy in the HN government.

Special Forces Approaches to Insurgent Strategy

3-43. Each insurgency is unique, calling for unique approaches. Direct approaches use a shape, clear, hold, build, and transition method. Indirect methods use negotiation as well as identify, separate, isolate, influence, and reintegration approaches. Insurgencies vary their strategies from urban or terror approaches to rural strategies that progressively deny government control. SF, through close contact with the population and as experienced advisors to security forces, develops a firsthand understanding of the situation and can develop access in troubled areas to influence conditions on the ground. With capabilities for lethal and nonlethal action, SF teams have the maturity to choose approaches and methods that lead toward an acceptable steady state, where grievances supporting the insurgency are resolved.

3-44. SF counters urban approaches by penetrating intelligence activities, locating enemy nets and cells, and conducting direct action. SF countered rural insurgent movements with several methods throughout history. Among these were the Civilian Irregular Defense Groups in Vietnam, the Joint Commission Observers in the Balkans, and Village Stability Operations in Afghanistan.

Village Stability Operations

The Village Stability Operations program personifies the U.S. Government's objectives and doctrine with the understanding of the counterinsurgency environment as it begins with understanding the population, then the insurgents, and finally the counterinsurgent. Living among the people, simultaneously focused on security, governance, and development with all efforts committed to protecting the populations at the village level, is one of the preventative measures that keeps the environment from being hostile.

(continued)

> **Village Stability Operations (continued)**
>
> For SF, this is a very comfortable line of operations. Since its inception, SF has been successful with this concept and the application of counterinsurgency and FID operations, to include living among the local populations within those countries when deployed. The Civilian Irregular Defense Groups in Vietnam and the Joint Commission Observers in the Balkans reflect operations that SF units have been conducting for the last 50 years.
>
> By building capacity and capability with the Afghan government and the Afghan people, this available resource—also known as the "through-and-with" concept—applies a process of developing the capability and capacity needed to meet the security objectives within a system that the local population finds palatable, as they are now the enforcers.
>
> Because of the immediacy and severity of security challenges found in rural conflict areas, such as Afghanistan, political and military reconsiderations of restrictions previously imposed upon some populations are driving centric counterinsurgency strategies. Civilian populations, identified as the centers of gravity in this insurgency, are normally found in the urban built-up areas. The bottom-up approach provides stability for rural villages within key districts and empowers the local government structure, which facilitates the village leadership to connect with the Government of the Islamic Republic of Afghanistan. As a major line of operations, Village Stability Operations transform an area of conflict as it develops a community base by providing a resistance to the insurgent's policies and practices, thus keeping the community out of the conflict.
>
> The Village Stability Operations program not only protects the population, but it enables the population to protect itself. Village Stability Operations blend formal and informal forms of governance into cohesive, trust building, and dynamic communities, and ensures that integrated systems of development, governance, and security are shared to establish stability. Village Stability Operations are an essential supporting component nested within the International Security Assistance Force operational objectives and the U.S. strategic military objectives.

Special Forces Regional Knowledge

3-45. SF Soldiers and their operations are particularly valuable during a counterinsurgency. These Soldiers have an appreciation of the essential nature and nuances of the conflict, and a close-up understanding of the motivation, strengths, and weaknesses of the insurgents and other actors located in their area of operations. This region-specific knowledge combined with their ability to gather special intelligence make SF Soldiers ideally suited to contribute to the overall counterinsurgency strategy. By using indigenous agencies and sources, the ODA can assemble and relay intelligence information to the SOTF and combatant commander, providing best estimates of the situation. This ground truth and the proximity of SF teams to the population offer commanders and planners with potent tools to counter insurgent tactics and strategies.

3-46. SF teams committed to counterinsurgency have a dual mission. They must assist the HN forces to defeat or neutralize the insurgent forces so that they may resume normal policing functions in what were once contested or insurgent-controlled areas. In addition, SF teams support the overall counterinsurgency program by conducting an advisory mission, influencing government agencies and forces to address grievances. Doing so provides an environment where the HN government can win the trust and support of its people and become self-sustaining. Both aspects of the counterinsurgency mission are of equal importance and must be conducted at the same time. A full treatment of the nature of counterinsurgency can be found in FM 3-24.

SECURITY FORCE ASSISTANCE

3-47. Security force assistance and FID overlap without being subsets of each other. JP 3-22 defines security force assistance as "the Department of Defense activities that contribute to unified action by the U.S. Government to support the development of the capacity and capability of foreign security forces and their supporting institutions."

3-48. Security force assistance is DOD's contribution to unified action to develop (organize, train, equip, rebuild, and advise) the capacity and capability of foreign security forces from the ministerial level down to units of those forces. Foreign security forces include, but are not limited to, the military; police; border police, coast guard, and customs officials; paramilitary forces; forces peculiar to specific nations, states, tribes, or ethnic groups; prison, correctional, and penal services; infrastructure-protection forces; and the governmental ministries and departments responsible for foreign security forces. At operational and strategic levels, both security force assistance and FID focus on developing the foreign security forces' internal capacity and capability. However, security force assistance also prepares foreign security forces to defend against external threats and to perform as part of an international coalition. FID and security force assistance are similar at the tactical level where advisory skills are applicable to both. SF teams performing security force assistance initially assess the foreign security forces they will assist and then establish a shared, continual way of assessing throughout development of the foreign security forces.

Organizing

3-49. Security force assistance includes organizing institutions and units, which can range from standing up a ministry to improving the organization of the smallest maneuver unit. Building capability and capacity in this area includes personnel, logistics, and intelligence and their support infrastructure. Developing HN tactical capabilities alone is inadequate; strategic and operational capabilities must be developed as well. HN organization and units should reflect their own requirements, interests, and capabilities—they should not simply mirror existing external institutions.

Training

3-50. Training occurs in training centers, academies, and units. Training includes a broad range of subject matter, to include security forces responding to civilian oversight and control.

Equipping

3-51. Equipping is accomplished through traditional security assistance, foreign support, and donations. Equipment must be appropriate for the HN's ability to sustain it. The equipment must also be appropriate for the physical environment of the region and the HN's ability to operate and maintain it.

Rebuilding

3-52. In many cases, particularly after major combat operations, it may be necessary to rebuild existing or build new infrastructure to support foreign security forces. This infrastructure includes facilities and materiel but may also include other infrastructure, such as command systems and transportation networks.

Advising

3-53. Advising HN units and institutions is essential to the ultimate success of security force assistance. Advising benefits both the state and the supporting external organizations. To successfully accomplish the security force assistance mission, advising requires specially trained ARSOF personnel who—

- Understand the operational environment.
- Provide effective leadership.
- Build legitimacy.
- Manage information.

Chapter 3

- Ensure unity of effort and unity of purpose.
- Can sustain the effort.

SPECIAL RECONNAISSANCE

3-54. JP 3-05 defines special reconnaissance as "reconnaissance and surveillance actions conducted as a special operation in hostile, denied, or politically sensitive environments to collect or verify information of strategic or operational significance, employing military capabilities not normally found in conventional forces." Special reconnaissance may include information on activities of an actual or potential enemy or secure data on the meteorological, hydrographic, or geographic characteristics of a particular area. Special reconnaissance may also include assessment of chemical, biological, residual nuclear, radiological, or environmental hazards in a denied area. Special reconnaissance includes target acquisition, area assessment, and post-strike reconnaissance.

3-55. Special reconnaissance complements national and joint operations area intelligence collection assets and systems by obtaining specific, well-defined, and time-sensitive information of strategic or operational significance. It may complement other collection methods constrained by weather, terrain-masking, or hostile countermeasures. Selected ARSOF conduct special reconnaissance as a human intelligence activity to place "eyes on target," when authorized, in hostile, denied, or diplomatically sensitive territory. Special reconnaissance typically provides essential information for a commander's situational awareness for a command decision, follow-on mission, or critical assessment.

3-56. In the operational environment, the SF and conventional command relationship may be that of supported and supporting, rather than tactical control or operational control. Use of SF with conventional forces by a JFC creates an additional capability to achieve objectives that may not be otherwise attainable. Using SF for special reconnaissance enables the JFC to take advantage of SOF core competencies to enhance situational awareness and facilitate staff planning of and training for unified action. However, such use does not mean that SF will become a dedicated reconnaissance asset for conventional forces. Instead, the JFC (through a joint special operations task force [JSOTF] or a TSOC) may task an SF element to provide special reconnaissance information to conventional forces that may be operating for a period of time within a joint special operations area or may task an SF element on a case-by-case basis to conduct special reconnaissance within a conventional force area of operations. In addition, SF and conventional elements working within the same area of operations may develop formal or informal information-sharing relationships that enhance each other's operational capabilities. SF may also employ advanced reconnaissance and surveillance sensors and collection methods that utilize indigenous assets.

DIRECT ACTION

3-57. *Direct action* is short-duration strikes and other small-scale offensive actions conducted as a special operation in hostile, denied, or politically sensitive environments and which employ specialized military capabilities to seize, destroy, capture, exploit, recover, or damage designated targets. Direct action differs from conventional offensive actions in the level of physical and political risk, operational techniques, and the degree of discriminate and precise use of force to achieve specific objectives (JP 3-05). In the conduct of these operations, SF may employ raid, ambush, or direct assault tactics (including close-quarters battle); emplace mines and other munitions; conduct standoff attacks by fire from air, ground, or maritime platforms; provide terminal guidance for precision-guided munitions; conduct independent sabotage; and conduct antiship operations.

3-58. Normally limited in scope and duration, direct action operations usually incorporate an immediate withdrawal from the planned objective area. These operations can provide specific, well-defined, and often time-sensitive results of strategic and operational critical significance.

3-59. SF may conduct direct action independently or as part of larger conventional or unconventional operations or campaigns. Although normally considered close combat, direct action also includes sniping

and other standoff attacks by fire delivered or directed by SF. Standoff attacks are preferred when the target can be damaged or destroyed without close combat. SF employs close combat tactics and techniques when the mission requires—
- Precise or discriminate use of force.
- Recovery or capture of personnel or materiel.

3-60. Direct action missions may also involve locating, recovering, and restoring to friendly control selected persons or materiel that are isolated and threatened in sensitive, denied, or contested areas. These missions usually result from situations that involve political sensitivity or military criticality of the personnel or materiel being recovered from remote or denied areas. These situations may arise from a political change, combat action, chance happening, or mechanical mishap. Direct action operations differ from personnel recovery by the use of—
- Dedicated ground combat elements.
- Unconventional techniques.
- Precise survivor-related intelligence.
- Indigenous assistance.

3-61. Direct action may be unilateral or combined actions, but are still short-duration, discrete actions. SF Soldiers execute direct action to achieve the supported commander's objectives. Additional information regarding the conduct of SF direct action operations can be found in FM 3-05.203.

COUNTERTERRORISM

3-62. JP 3-26 defines counterterrorism as "actions taken directly against terrorist networks and indirectly to influence and render global and regional environments inhospitable to terrorist networks." SF units possess the capability to conduct these operations in environments that may be denied to conventional forces because of political or threat conditions.

3-63. HN responsibilities, Department of Justice and Department of State lead agency authority, legal and political restrictions, and appropriate DOD directives limit SF involvement in counterterrorism. The role and added capability of SF is to conduct offensive measures within the overall combating terrorism efforts of the DOD. SF units conduct counterterrorism missions as special operations by covert, clandestine, or low-visibility means. SF activities within counterterrorism include—
- Intelligence operations to collect, exploit, and report information on terrorist organizations, personnel, assets, and activities. SF has the capability to conduct these operations in an overt, covert, or clandestine manner.
- Network and infrastructure attacks to execute preemptive strikes against terrorist organizations. The objective is to destroy, disorganize, or disarm terrorist organizations before they can strike targets of national interest.
- Hostage or sensitive materiel recovery to rescue hostages or to recover sensitive materiel from terrorist control. These activities require capabilities not normally found in conventional military units. Ensuring the safety of the hostages and preventing destruction of the sensitive materiel are essential mission requirements.
- Nonlethal activities to defeat the ideologies or motivations that spawn terrorism by nonlethal means. These activities could include FID and a range of information-related capabilities integrated though information operations.

Note: Most counterterrorism activities are classified. Further discussion of counterterrorism is beyond the scope of this publication.

COUNTERPROLIFERATION OF WEAPONS OF MASS DESTRUCTION

3-64. JP 3-40 defines counterproliferation as "those actions taken to defeat the threat and/or use of weapons of mass destruction against the United States, our forces, allies, and partners." JP 3-40 defines weapons of mass destruction as "chemical, biological, radiological, or nuclear weapons capable of a high order of destruction or causing mass casualties and exclude the means of transporting or propelling the weapon where such means is a separable and divisible part from the weapon." The major objectives of combating weapons of mass destruction policy, which include nonproliferation, counterproliferation, and consequence management activities, are to prevent the acquisition of weapons of mass destruction and delivery systems, to roll back proliferation where it has occurred, to deter and defeat the use of weapons of mass destruction and their delivery systems, to adapt U.S. military forces and planning to operate against the threats posed by weapons of mass destruction and their delivery systems, and to mitigate the effects of weapons of mass destruction use. The continued spread of weapons of mass destruction technology can foster regional unrest and provide terrorist organizations with new and potent weapons. SF provides the following capabilities for this core activity:

- Expertise, materiel, and teams to supported combatant command teams to locate, tag, and track weapons of mass destruction, as required.
- Capabilities to conduct direct action in limited access areas, as required.
- Partnership-building capacity for conducting counterproliferation activities.
- Capabilities to coordinate the employment of information-related capabilities integrated through information operations.

Note: Specific counterproliferation activities conducted by SF are classified. Further discussion of counterproliferation is beyond the scope of this publication.

PREPARATION OF THE ENVIRONMENT

3-65. SF conducts preparation of the environment as a type of shaping activity supporting the other principal tasks that may be conducted in the future. Preparation of the environment is an umbrella term for actions taken to prepare the operational environment for potential operations. Preparation of the environment consists of operational preparation of the environment, advanced force operations, and intelligence operations (USSOCOM Directive 525-16).

Operational Preparation of the Environment

3-66. Operational preparation of the environment is the conduct of activities in likely or potential areas of operations to prepare and shape the operational environment. Combatant commanders conduct operational preparation of the environment to develop knowledge of the operational environment, establish human and physical infrastructure, and for general target development. Operational preparation of the environment activities include passive observation, area familiarization, site surveys, mapping the information environment, military source operations, developing nonconventional assisted recovery capabilities, use of couriers, developing safe houses and assembly areas, positioning transportation assets, and cache emplacement and recovery (USSOCOM Directive 525-16).

Advanced Force Operations

3-67. Advanced force operations are conducted to refine the location of specific, identified targets and further develop the operational environment. Advanced force operations encompass many operational preparation of the environment activities, but they are intended to prepare for near-term direct action. Advanced force operations include close target reconnaissance; tagging, tracking, and locating; reception, staging, onward movement, and integration of forces; infrastructure development; and terminal guidance. Unless specifically withheld, advanced force operations also include direct action in situations when failure to act means loss of a fleeting opportunity for success (USSOCOM Directive 525-16).

Intelligence Operations

3-68. Intelligence operations include human intelligence activities (including military source operations); counterintelligence activities; airborne, maritime, and ground-based signals intelligence; tagging, tracking, and locating; and intelligence, surveillance, and reconnaissance. Intelligence operations complement operational preparation of the environment and advanced force operations in the overall preparation of the environment (USSOCOM Directive 525-16).

3-69. SF conducts preparation of the environment in support of GCC plans and orders to alter or shape the operational environment to create conditions conducive to success across the range of military operations. Preparation of the environment sets the conditions for operational success by cultivating relationships, establishing networks of partners, and laying the groundwork that will facilitate the conduct and sustainment of future operations. SF routinely shapes the operational environment and gains access to nations through its inherent activities, operations, actions, and tasks. Through the regional initiative of shaping, SF actively seeks to develop or improve regional influence and the ability to conduct follow-on operations. The regional focus, cross-cultural insights, language capabilities, and relationships of SOF provide access to and influence nations where the presence of conventional U.S. forces is not warranted.

This page intentionally left blank.

Chapter 4
Organization

Do not try to do too much with your own hands. Better the Arabs do it tolerably than that you do it perfectly. It is their war, and you are to help them, not to win it for them. Actually, also, under the very odd conditions of Arabia, your practical work will not be as good as, perhaps, you think it is.

T.E. Lawrence
27 Articles

Throughout history, success by a small force against a strategic or operational objective required combinations of special equipment, training, people, or tactics that go beyond those found in conventional units. These formations and tactics made possible unconventional approaches in an operating environment ill-suited to conventional military solutions and formations. This chapter describes the authorities and responsibilities of SF, the formal table of organization and equipment, and the mission, organization, function, and characteristics of the SFG and its organic units.

UNITED STATES SPECIAL OPERATIONS COMMAND

4-1. The USSOCOM's components are the USASOC, the Naval Special Warfare Command, the Air Force Special Operations Command, and the Marine Corps Special Operations Command (Figure 4-1). The Joint Special Operations Command is a USSOCOM subunified command. The Commander, USASOC, is designated the Army Service Component Commander of USSOCOM. The Army component is under the combatant command of the Commander, USSOCOM. Additionally, USASOC headquarters is designated the USSOCOM alternate command headquarters.

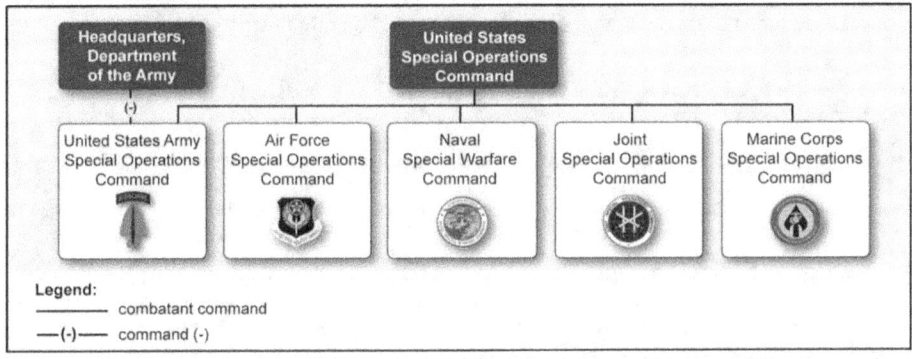

Figure 4-1. United States Special Operations Command organization

UNITED STATES ARMY SPECIAL OPERATIONS COMMAND

4-2. USASOC provides SF, Ranger, special operations aviation, MIS, and CA personnel to GCCs and U.S. Ambassadors. Commander, USASOC, exercises command of continental United States-based Regular ARSOF. Commander, USASOC, is responsible for the organization, administration, recruiting, equipping,

Chapter 4

training, maintenance, support, readiness, deployment, and education of assigned forces, including those forces temporarily assigned to the operational control of other unified commanders.

4-3. USASOC consists of four component subordinate commands and three component subordinate units manned with civilians and Regular Army and Reserve Component military personnel (Figure 4-2). The component subordinate commands of USASOC are the USASFC, the U.S. Army John F. Kennedy Special Warfare Center and School (USAJFKSWCS), the Army Special Operations Aviation Command, and the Military Information Support Operations Command. The component subordinate units are the 95th Civil Affairs Brigade, the 75th Ranger Regiment, and the 528th Sustainment Brigade (Special Operations).

Figure 4-2. United States Army Special Operations Command organization

UNITED STATES ARMY JOHN F. KENNEDY SPECIAL WARFARE CENTER AND SCHOOL

4-4. USAJFKSWCS, the U.S. Army's Special Operations Center of Excellence, trains, educates, develops, and manages world-class CA, Psychological Operations, and SF warriors and leaders in order to provide the ARSOF regiments with professionally trained, highly educated, innovative, and adaptive operators. As a component subordinate command, the USAJFKSWCS serves as the USASOC proponent for all matters pertaining to individual training, develops doctrine and all related individual and collective training material, provides leader development, develops and maintains the proponent training programs and systems, and provides entry-level and advanced individual training and education for CA, MIS, and SF. A complete description of this organization and its mission is contained in FM 3-05.

UNITED STATES ARMY SPECIAL FORCES COMMAND

4-5. The mission of the USASFC is to organize, equip, train, validate, and deploy forces to conduct special operations across the range of military operations, in support of the USSOCOM, regional combatant commanders, U.S. Ambassadors, and other government agencies, as directed. The USASFC comprises five Regular Army SFGs and two Army National Guard SFGs (Figure 4-3, page 4-3).

Organization

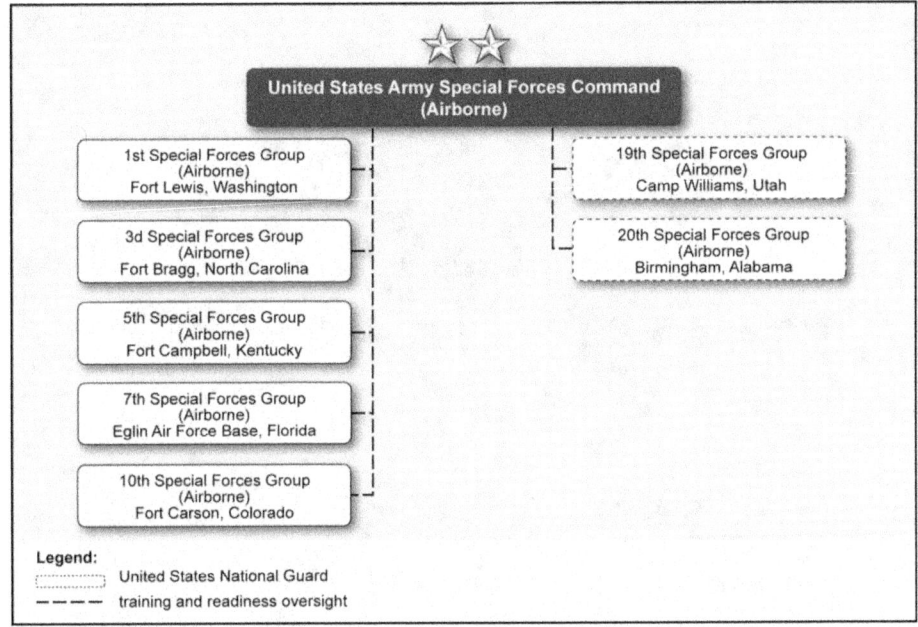

Figure 4-3. United States Army Special Forces Command (Airborne) organization

4-6. Regional orientation is the hallmark of SF Soldiers and units, with each of the five Regular Army and two Army National Guard SFGs regionally aligned with a GCC (Figure 4-4, page 4-4). Because of this regional alignment, senior SF Soldiers may sometimes have years of significant experience operating in a particular region. Combined with language capabilities that support operations within these regions, SF Soldiers are able to develop long-term relationships with indigenous personnel that provide GCCs, U.S. Ambassadors, and conventional forces critical capabilities and knowledge concerning specific countries and areas. Commanders at all levels within SF will routinely focus and incorporate regional orientation in terms of language, environment, and cultural uniqueness into different aspects of training.

SPECIAL FORCES GROUP

4-7. The SFG is the largest combat element of ARSOF. The SFG is extremely flexible and capable of forming the nucleus of a JSOTF and is normally identified as the Army service force component of a JSOTF. Each SFG is regionally aligned with one or more geographic combatant commands.

Mission

4-8. The mission of the SFG is to plan, execute mission command, and support special operations activities in any operational environment. Skills required for special operations—combined with the quality, motivation, and experience found in the SFG—enable it to perform a multitude of missions. The SFG is self-contained with organic mission command and support elements. For long-duration missions or large-scale operations, further augmentation by either joint or conventional forces is required.

Chapter 4

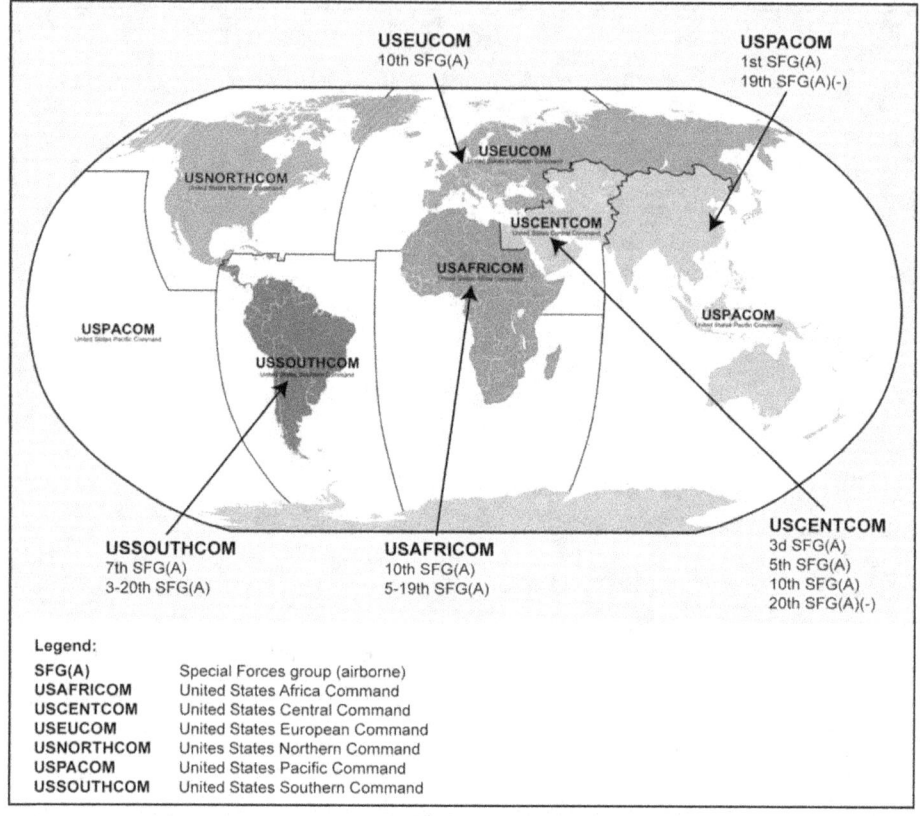

Figure 4-4. Regional orientation of the Special Forces groups

Organization

4-9. The SFG consists of an HHC, a group support battalion, and four SF battalions (Figure 4-5, page 4-5).

Function

4-10. SFGs execute special operations in support of regional engagements as directed by the GCCs, Joint Chiefs of Staff exercises, and other deployments for training. They are the force of great utility to the JFC, especially when a low-profile military solution is the most politically acceptable. During conflict, SF functions best as a force multiplier, but it is capable of unilateral direct action or special reconnaissance in support of major combat operations.

4-11. The SFG headquarters commands and controls assigned and attached forces. It plans, coordinates, and directs SF operations separately or as a part of a larger force. It also—

- Provides command and staff personnel to establish and operate a JSOTF when augmented by resources from other Services.
- Functions as the headquarters for a SOTF.

Organization

- Directs the activities of up to four battalion-sized SOTFs.

Note: A JSOTF normally plans to operate only four SOTFs according to the table of organization and equipment; however, if a situation dictates, a JSOTF can direct the operations of as many SOTFs as are required within the area of responsibility. The additional SOTFs can be fielded from either uncommitted SFGs or other joint SOF.

- Provides support for the sustainment of the activities of deployed ODAs.
- Advises, coordinates, and assists the staff on employing SF elements to a TSOC, JSOTF, security assistance organization, or other major headquarters.
- Provides cryptographic material support to the SFG and its subordinate elements.

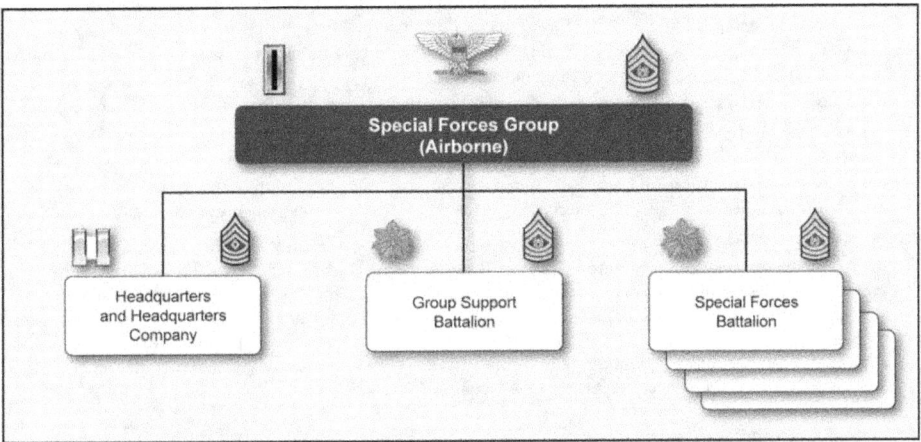

Figure 4-5. Special Forces group (airborne) organization

Characteristics

4-12. The SFG is designed to be a versatile, self-contained organization that can provide a TSOC or the GCC with an extremely flexible force capable of operating in ambiguous and swiftly changing scenarios. Among their capabilities, the SFG can—

- Infiltrate and exfiltrate an operational area by air, land, or sea.
- Develop, organize, equip, train, and advise or direct indigenous military or paramilitary forces.
- Plan and conduct unilateral SF principal tasks.
- Train, advise, and assist other U.S. and coalition forces or agencies.
- Perform other special operations, as directed.

4-13. The following factors need to be taken into account whenever the employment of SF is considered:

- The SFG is not a substitute for conventional forces. The SFG is neither trained nor equipped to conduct sustained conventional combat operations, and therefore should not be substituted for conventional units that are able to effectively execute that mission.
- The SFG has no organic aviation assets, although there are aviation officers as part of the group staff who function as mission planners.

Chapter 4

- The SFG is not a combined arms organization. It cannot conduct conventional combined arms operations on a unilateral basis. The SFG abilities are limited to advising or directing indigenous military forces or conducting these types of operations in conjunction with conventional or other joint forces.
- SFGs cannot maintain themselves for extended periods without significant sustainment support from a conventional support structure. The SFG will depend on the resources of the theater army to support and sustain long-term operations.

Headquarters and Headquarters Company

4-14. The group HHC provides routine administrative and logistics support to the group headquarters. Figure 4-6, page 4-7, shows the group HHC organization.

4-15. The group headquarters commands and controls assigned and attached forces. It plans, coordinates, and directs SF operations separately or as a part of a larger force. It also—

- Can execute mission command from a forward support base in country, or in an adjacent or nearby country within the same theater.
- Exercises command and staff personnel to establish and operate a SOTF, JSOTF, or combined joint special operations task force (CJSOTF).
- Directs the activities of subordinate units.
- Provides support for the sustainment of deployed subordinate units.
- Provides cryptographic material support to the SFG and its subordinate elements.

4-16. The company headquarters provides routine administrative and logistics support to the group headquarters. It depends on the group support battalion for unit-level maintenance of its organic wheeled vehicles, power-generation equipment, and signal equipment. When the group establishes a JSOTF/CJSOTF, the HHC commander serves as headquarters commandant under the direct supervision of the deputy group commander. As headquarters commandant, the HHC commander is responsible for the movement, internal base operations and administration (including space allocation, billeting, and food service), defense, and physical security of the JSOTF/CJSOTF.

Chemical Reconnaissance Detachment

4-17. The chemical reconnaissance detachment is a USASFC asset attached to an SFG. The chemical reconnaissance detachment may be task organized within the group to satisfy mission requirements. These special detachments are the only chemical reconnaissance detachments with this mission in the U.S. Army. USASFC currently has two Regular Army chemical reconnaissance detachments, three U.S. Army Reserve chemical reconnaissance detachments, and two Army National Guard chemical reconnaissance detachments.

4-18. The chemical reconnaissance detachment conducts chemical, biological, radiological, nuclear, and high-yield explosives (CBRNE) reconnaissance and sampling in permissive or uncertain environments. It supports the requirements of the SFG commander, SOF commanders, and GCCs at the strategic, operational, and tactical levels.

4-19. The chemical reconnaissance detachment commander is a captain and the detachment sergeant is a master sergeant. These two individuals make up the detachment headquarters section. The four internal chemical detachments are composed of four chemical operations noncommissioned officers of various ranks. Each position within the unit has its own operations and functions. The chemical reconnaissance detachment is capable of supporting all CBRNE aspects of SF missions. The team can augment an ODA to perform tasks involved in detecting, collecting, packaging, and identifying CBRNE material. The team can conduct its mission unilaterally in permissive environments and in an uncertain environment when accompanied by an ODA. Having four noncommissioned officers per chemical detachment allows the detachments to conduct split-team operations when the situation does not warrant a full team. The chemical reconnaissance detachment can serve as a manpower pool from which SOF commanders at all levels can organize a tailored composite team to perform a specific mission.

Organization

4-20. The chemical reconnaissance detachment can identify potential hazards and can confirm or deny the use of CBRNE in proximity to the JSOTF or SOTF. Additionally, the chemical reconnaissance detachment can train SOF and foreign personnel in both individual and collective CBRNE tasks.

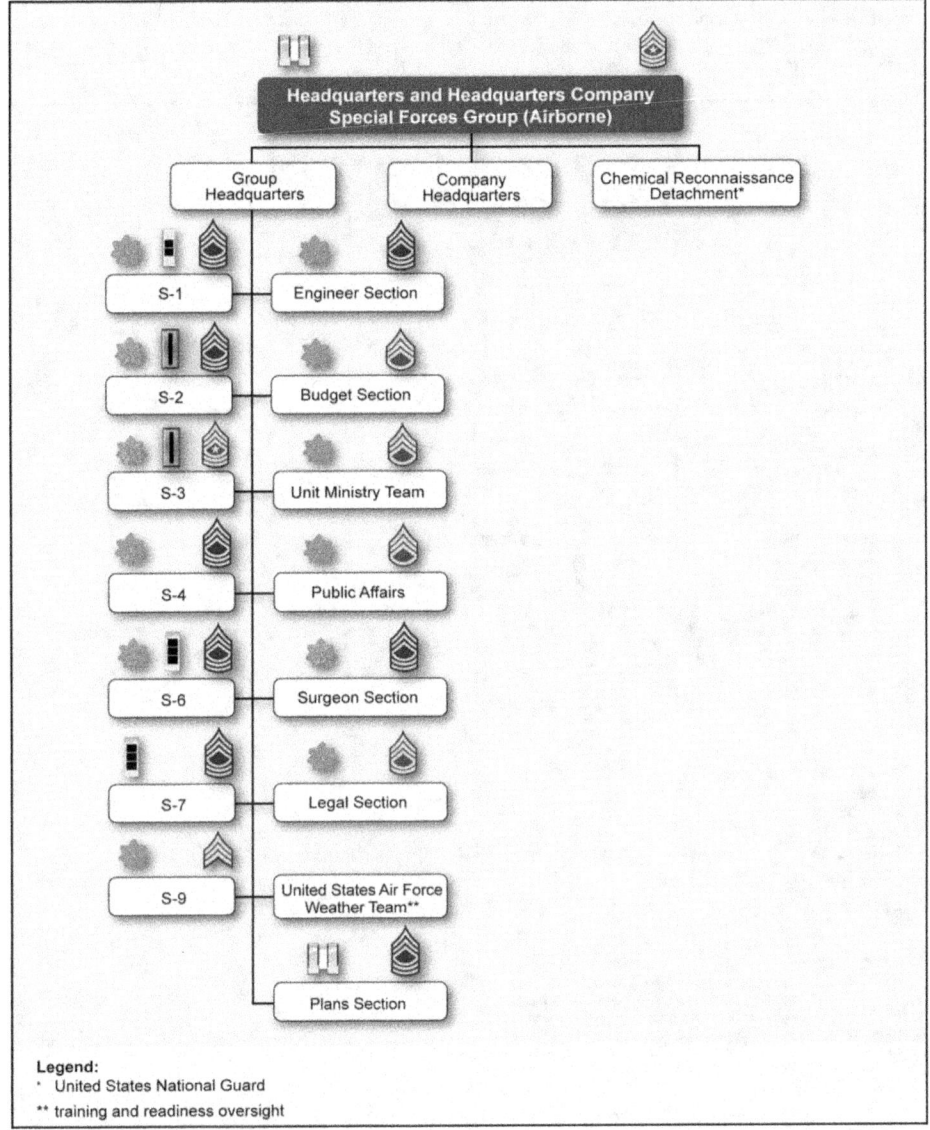

Figure 4-6. Headquarters and headquarters company (airborne) organization

Chapter 4

Group Support Battalion

4-21. The group support battalion concept (Figure 4-7) consists of a headquarters and headquarters detachment (HHD), a supply and distribution company, a maintenance company, a medical company, and three forward support companies. The group support battalion ties together the entire sustainment spectrum of supplies, maintenance, and services. The battalion commander is the SFG commander's senior battle logistician. The group support battalion plans, coordinates, executes logistic sustainment operations and, when directed, supports the forces attached or assigned to the JSOTF.

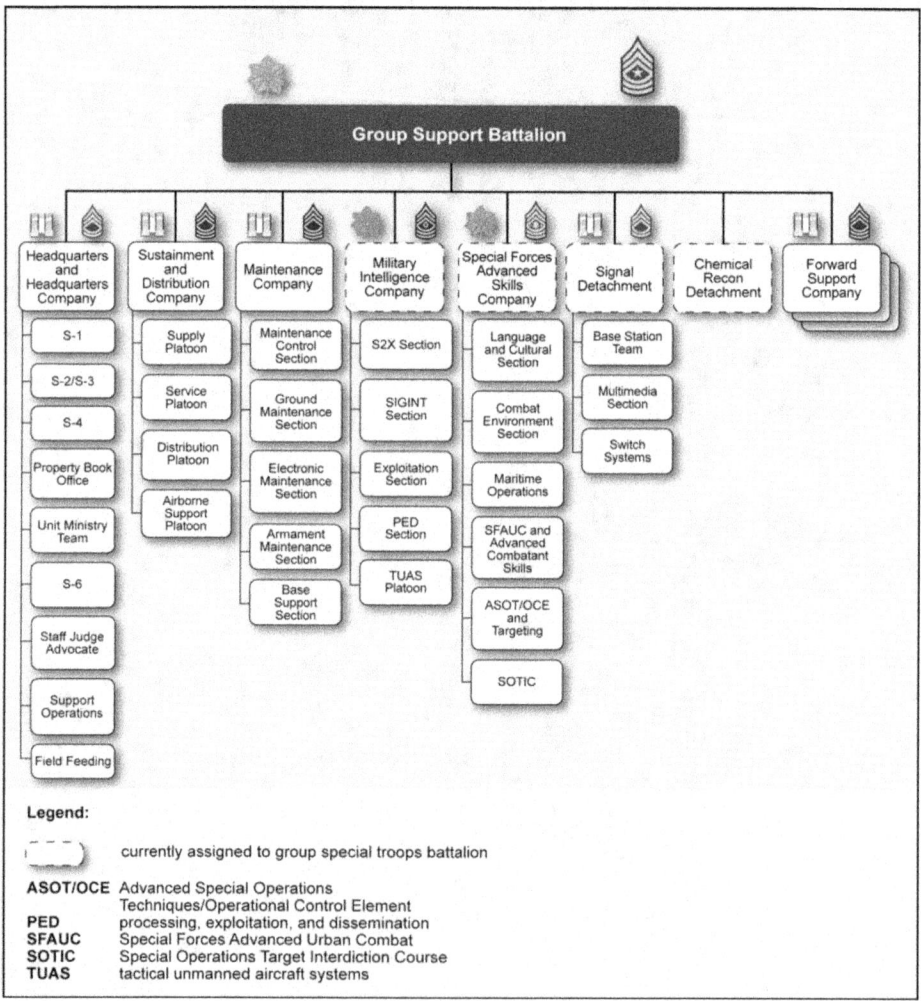

Figure 4-7. Group support battalion (airborne) organization

Forward Support Company

4-22. The forward support company provides routine administrative and logistics support to the SF battalions. The forward support company comprises the sustainment, distribution, and maintenance platoons. The forward support company is a multifunctional logistics company providing maintenance, limited Class I through Class IX supplies, fuel and water production, ammunition holding, and transportation. The forward support company is capable of providing for the entire SF battalion and its attached elements. When the SF battalion establishes a SOTF, the forward support company commander may coordinate and supervise the support center logistics activities.

Group Special Troops Battalion

4-23. The special troops battalion provides special operational and training support for the SFG. Proposed changes will meld these support units into the SFG support battalion and 4th battalion. The special troops battalion now consists of a command section, an SF advanced skills company, a chemical decontamination detachment, a signal detachment, and a military intelligence company, an information and dominance company, and a technical support company. The group special troops battalion is expected to be absorbed by the group support battalion.

Advanced Skills Company

4-24. The group advanced skills company is primarily responsible for training group-assigned elements or other specified personnel in select SF skill sets. The company contains a command section and the following elements:

- **Language and Cultural Section.** This section manages the group's language lab and obtains contract support for short-duration language training, as necessary. This section maintains and updates area handbooks and cultural information to support ODA missions, and provides or contracts for translation support of documents and training material. When formed as the isolation facility, Soldiers from this section serve as liaison officers to isolated operational detachments.
- **Combat Environmental Section.** This section is responsible for researching, acquiring, and maintaining group area of responsibility-specific limited-use equipment that is not issued to ODAs because of prohibitive cost. This equipment includes items such as ground mobility vehicles, ski and snow equipment, snowmobiles, all-terrain vehicles, extreme cold-weather suits, and pack animal equipment. This section maintains subject-matter expertise on the use and maintenance of the equipment as well as the ability to provide training in the proper use of the equipment to ODAs during isolation and future operations. This section is also responsible for unit sustainment and training in survival and evasion tactics, techniques, and procedures, as well as the conduct of a survival, evasion, resistance, and escape level B program. When formed as the isolation facility, this section serves as liaison officers to the isolated ODAs.
- **Maritime Operations Section.** This section provides accountability and maintenance of all equipment used in the conduct of scuba, underwater breathing apparatus, and small-boat operations in accordance with applicable policies and regulations. All SF Soldiers assigned to this section will be qualified combat divers on active dive status. When formed as the isolation facility, this section facilitates the training and rehearsals of isolated ODAs.
- **Special Forces Advanced Urban Combat and Advanced Combat Skills Section.** This section trains ODAs and SF companies in SF advanced urban combat tactics, techniques, and procedures on a rotational basis. This section may be directed by the commander to train personnel in other advanced combat skills, such as advanced marksmanship training or unarmed combat. When formed as the isolation facility, this section in conjunction with the operations detachment command section will form the isolation facility operations cell and support cells.
- **Advanced Special Operations Techniques and Advanced Target Section.** This section is responsible for training group personnel and ODAs in the use of SF advanced special operations techniques. This section also conducts operations as directed by the group commander. This section is responsible for conducting liaison as necessary to support group training and operational initiatives.

Chapter 4

- **Special Forces Sniper Course Section.** This section is responsible for training group-assigned Soldiers in SF sniper course tactics, techniques, and procedures on a rotational basis, to include conducting refresher training for group-assigned Soldiers previously qualified as special operations snipers. This section may also be tasked to conduct advanced marksmanship training. When formed as the isolation facility, this section facilitates the training and rehearsals of isolated ODAs.
- **Special Operations Forces Multipurpose Canine Section.** The SOF multipurpose canine section provides a handler and a SOF multipurpose canine capable for use in the scout dog role, mine and tunnel exploitation and search, rural and urban combat tracking operations, off-leash site exploitation for cache and spider hole discovery, building searches, area searches, and clearing operations with an assault element. Additionally, the SOF multipurpose canine can detect explosive material and improvised explosive devices, as well as track, contain, and, when necessary, attack enemy personnel fleeing from targets or posing a threat to friendly forces. The SOF multipurpose canine can also perform suicide bomber alert and interdiction and suspect/insurgent apprehension and detention.

Chemical Decontamination Detachment

4-25. The group chemical decontamination detachment provides mission command and CBRNE support to the SFG in the form of CBRNE decontamination teams. The CBRNE decontamination headquarters and teams are consolidated at the group support company for administrative and training purposes. Four CBRNE decontamination teams provide CBRNE decontamination support to the JSOTF. When the battalions within the group deploy, these decontamination teams will also deploy to provide support for the SOTFs and AOBs.

Signal Detachment

4-26. The group signal detachment has two primary functions. It installs, operates, and maintains secure JSOTF communications with the SOTFs and the deployed ODBs and ODAs under the group's direct mission command. It also installs, operates, and maintains continuous internal JSOTF communications. This base communications support includes message center services, internal telephone communications, information management operations, photographic support, and electronic maintenance. When the group establishes a JSOTF, the signal detachment commander serves as the systems control officer. When the detachment is formally detached from the support company, the detachment commander exercises normal company-level command; however, the detachment depends on the support company for administrative and sustainment support. FM 3-05.160 provides additional information regarding signal organizations, communications, capabilities, and equipment within SF organizations.

Military Intelligence Company

4-27. The group military intelligence company contains most of the group's single-source and all-source analysis capability. The military intelligence company is responsible for collection management, all-source fusion of single-source information, analysis and production, dissemination of finished intelligence and geospatial products, generation of the geospatial data background for the common operational picture, and the control and management of the sensitive compartmented information communications team. The tactical unmanned aircraft system platoon falls under the military intelligence company.

Tactical Unmanned Aircraft System Platoon

4-28. The group tactical unmanned aircraft system platoon is designed to provide the SFG commander with a primary day and night reconnaissance, surveillance, and target acquisition system. The tactical unmanned aircraft system platoon provides enhanced situational awareness, target acquisition, and battle damage assessment and management, as well as the ability to penetrate into denied areas. The unmanned aircraft system provides the tactical maneuver commander near-real-time reconnaissance, surveillance, target acquisition, and protection, day or night and in limited adverse weather conditions. More detailed information regarding the unmanned aircraft system capabilities in the SFGs is in FM 3-76.

Organization

SPECIAL FORCES BATTALION

4-29. The SF battalion is similar in composition to the multipurpose, flexible warfighting organization that is the SFG. The SF battalion is normally identified for and has the capability to form the nucleus of a SOTF within a CJSOTF. The SF battalion will have a more direct role in the mission command, support, and sustainment of its own organic or attached ODAs and ODBs.

Mission

4-30. The SF battalion of the SFG plans, conducts, and supports special operations activities in any operational environment—permissive, uncertain, or hostile.

Organization

4-31. SF battalions consist of an HHD, three SF companies, a battalion support company, and, when deployed, an attached forward support company. Figure 4-8 shows the SF battalion organization.

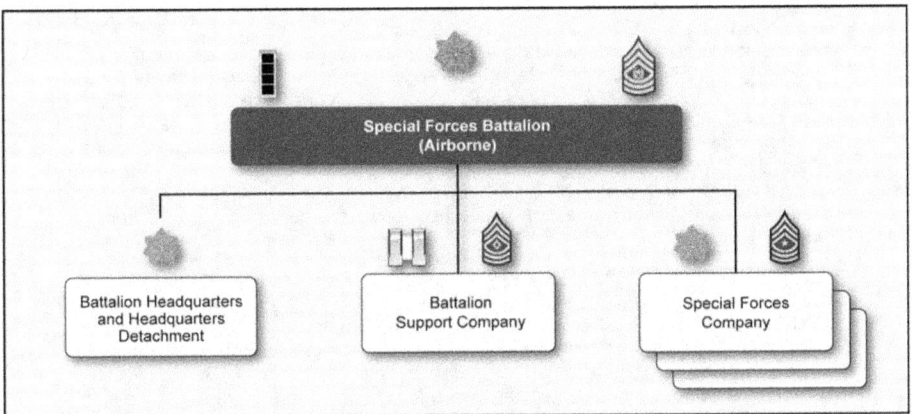

Figure 4-8. Special Forces battalion (airborne) organization

Function

4-32. Each SF operational detachment has its own multidigit number for identification. The numbering system reflects the adding of a fourth SF battalion to each of the five Regular Army SFGs. Although the two Army National Guard SFGs did not receive a fourth SF battalion, their numbering system is identical. A SOTF is identified by two digits; the first denotes its SFG and the second denotes the SF battalion within that SFG. For example, SOTF 12 identifies the 1st SFG and then the 2d Battalion. ODBs and ODAs are identified by four digits. The first two digits are identical to the SOTF. The third digit identifies the SF company within that SF battalion (1 denotes A Company, 2 denotes B Company, and so on). The fourth digit identifies the individual detachment within that SF company. All ODBs are identified by a 0 as the fourth digit. For example, ODB 3120 identifies the 3d SFG, 1st Battalion, B Company Headquarters. ODA 5432 identifies the 5th SFG, 4th Battalion, C Company, ODA number 2.

4-33. The battalion is directly responsible for isolating, deploying, controlling, sustaining, recovering, and reconstituting its assigned or attached ODBs and ODAs. When deployed, it has the capability to form a SOTF that is task organized to mirror the group functionalities.

Characteristics

4-34. The SF battalion has many of the same characteristics as the SFG. The SF battalion is capable of—
- Being rapidly deployed.
- Maintaining worldwide communications.
- Conducting early-entry operations, including infiltration into an operational area by airborne, air assault, or amphibious operations.
- Conducting operations in varying terrain, such as woodland, jungle, mountain, or desert environments.

The SF battalion must rely on the SFG for support in areas such as aerial delivery, electronic maintenance, multimedia, technical and information support, combat tracking, advanced skills training, and chemical reconnaissance.

Operation FOCUS RELIEF III

Operation FOCUS RELIEF III took place in Nigeria beginning in September 2001. The mission for this operation was to support United Nations peacekeeping efforts in the Republic of Sierra Leone by preparing elements of the Nigerian Army to deploy to Sierra Leone to conduct peacekeeping operations in the war-torn West African nation. The 3d Battalion of the 3d SFG formed Forward Operations Base 33 (what would now be referred to as SOTF 33), and deployed from Fort Bragg, North Carolina, to the Nigerian capital of Abuja. Once in Abuja, Forward Operations Base 33 came under the operational control of European Command and was charged with providing mission command for its organic and attached forces as they conducted the training mission.

Each of the three SF companies in the battalion also deployed to Nigeria and established an AOB in three different locations in the country: AOB 370 in Ilorin, AOB 380 in Serti, and AOB 390 in Bernin Kebbi. The three AOBs were partnered with three Nigerian Army battalions. The missions for the AOBs were to mission command their own organic ODAs and attachments as the ODAs trained the Nigerian battalions for peacekeeping operations. Following 12 weeks of training, Forward Operations Base 33 redeployed to the United States and each of the Nigerian Army battalions deployed for yearlong peacekeeping duties to the Republic of Sierra Leone.

Battalion Headquarters and Headquarters Detachment

4-35. The battalion HHD (Figure 4-9, page 4-13) mission commands the battalion and attachments. When deployed, it commands and controls the activities of a SOTF. It also trains and prepares subordinate units for deployment, and it directs, supports, and sustains the activities of deployed units.

Organization

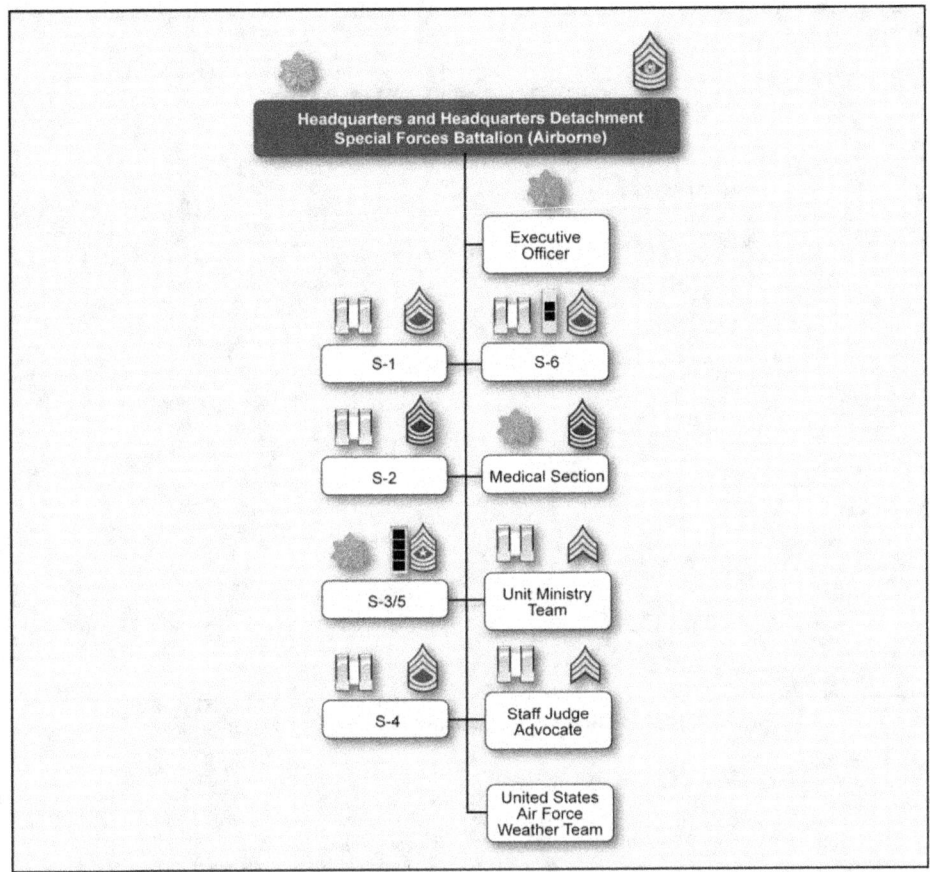

Figure 4-9. Battalion headquarters and headquarters detachment (airborne) organization

Battalion Support Company

4-36. The battalion support company consists of a company headquarters and two detachments (signal and military intelligence) and provides routine administrative and logistics support to the battalion headquarters detachment, the support company's organic elements, and the battalion's SF companies (Figure 4-10, page 4-14). The battalion support company commander commands all personnel and elements assigned or attached to the company. When the battalion establishes a SOTF, the battalion support company commander normally serves as the support center director. The support center is discussed in further detail beginning in Chapter 5. The support center director's duties require direct interface with the group support battalion or battalion and Army Service component command logistics support elements. In coordination with the operations staff officer and headquarters commandant, the battalion support company commander prepares the base defense plan and supervises the activities of the base defense operations cell. When all ODBs are committed to other missions, the support center director provides support to all the uncommitted ODAs and attached special operations teams A (SOT-As) at the SOTF, and supervises their pre-mission training activities in coordination with the operations center, which is discussed in Chapter 5.

Chapter 4

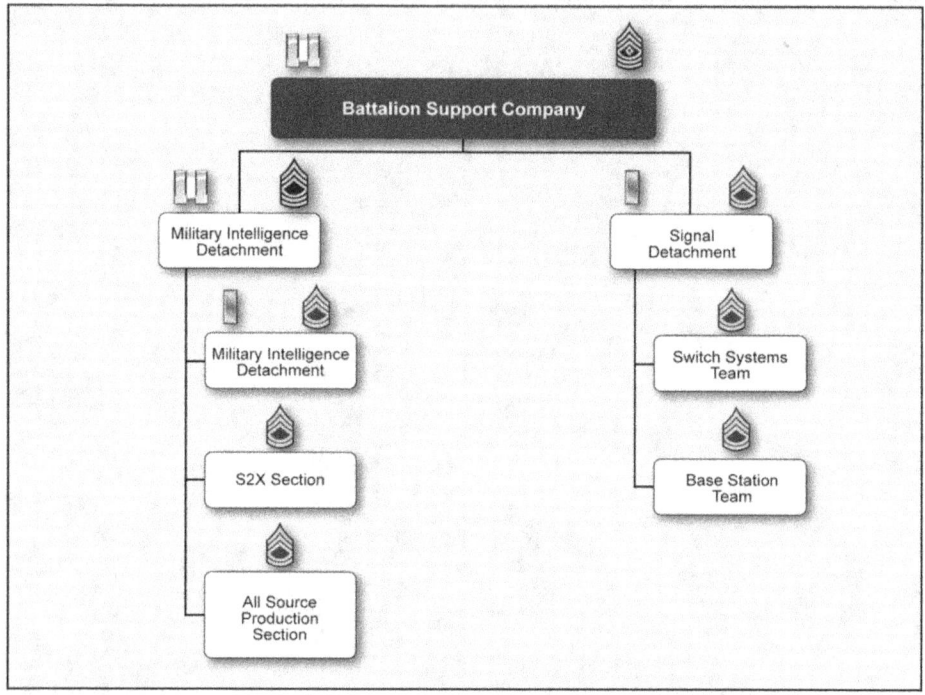

Figure 4-10. Battalion support company (airborne) organization

Forward Support Company

4-37. The forward support company performs unit-level supply, service, distribution, and maintenance functions for the entire SF battalion and its attached elements. When the battalion establishes a SOTF, the forward support company commander coordinates and supervises support center logistics activities and works for the battalion support company commander under the staff supervision of the battalion logistics staff officer. The forward support company does not have any organic truck drivers in its supply and transportation section.

Battalion Signal Detachment

4-38. The battalion signal detachment has two primary functions. It installs, operates, and maintains secure external communications with higher or adjacent SOF and conventional headquarters and its deployed ODBs and ODAs. It also installs, operates, and maintains internal battalion communications. This base communications support includes message center services, internal telephone communications, information management operations, and electronic maintenance. The detachment has no organic multimedia or communications security section; however, the signal detachment normally maintains its own communications security subaccount.

4-39. When the SF battalion deploys and establishes a SOTF, the signal detachment commander serves as the signal center systems control officer and assistant signal center director. When the signal detachment is formally detached from the support company, the detachment commander exercises normal company-level command. The signal detachment, however, depends on the support company for administrative and logistics support.

Organization

Battalion Military Intelligence Detachment

4-40. The military intelligence detachment provides intelligence support to SF battalion or below operations based on mission requirements. The detachment consists of the following intelligence disciplines: signals intelligence, human intelligence, imagery intelligence, counterintelligence, and all-source intelligence analysts.

4-41. When in direct support to SOTF operations, the all-source analysts will be detached to the intelligence staff section. Additionally, human intelligence and counterintelligence assets will be detached under direction of the SOTF commander to ensure protection from espionage, as well as to augment source operations infrastructure.

4-42. The tactical control and analysis element/special operations team B (SOT-B) section provides signals intelligence support to the SOTF, and serves as the primary conduit for signals intelligence reporting to elements above the SOTF. The tactical control and analysis element/SOT-B terminates signals intelligence reporting from the SOT-As and provides 24-hour operation. The tactical control and analysis element/SOT-B also provides a diverse language pool to support SF operations and provides signals intelligence support to deployed AOBs acting independently or as a special operations command and control element (SOCCE).

4-43. SOT-As are low-level signals intelligence collection teams that intercept and report operational and technical information derived from tactical threat communications through prescribed communications paths. The mission of a SOT-A is to conduct signals intelligence and electronic warfare in support of cyber/electromagnetic activities, unilaterally or in conjunction with other SOF elements to support existing and emerging SOF missions.

4-44. SOT-As perform a number of different mission and collateral activities in support of larger special operations. The primary roles of SOT-As include—

- Electronic reconnaissance.
- Protection.
- Signals research and target development.

4-45. Collateral activities include—

- Support to personnel recovery.
- Support to other special operations.

SPECIAL FORCES COMPANY

4-46. A significant difference exists between most conventional Army companies and SF companies. An SF company headquarters is an ODB, which is commanded by an experienced SF major. The composition of an SF company, which is described in further detail below, enables the SF company to mission command its own organic or attached ODAs in garrison and when deployed. It also provides the battalion commander with an additional operational element that can conduct its own assigned mission. The mission may require the ODB to operate separately or exercise mission command of a mix of organic and/or attached ODAs.

Mission

4-47. The SF company (Figure 4-11, page 4-16) plans and conducts special operations activities in any operational environment—permissive, uncertain, or hostile—and can deploy to conduct SF operations in a specified area of operations or joint special operations area. In this capacity, the ODB can plan and conduct SF operations either separately or as part of a larger force. The ODB can also provide operational support to either the SF battalion of an SFG by establishing and operating a SOCCE or an SF liaison element, an AOB, or an isolation facility.

Chapter 4

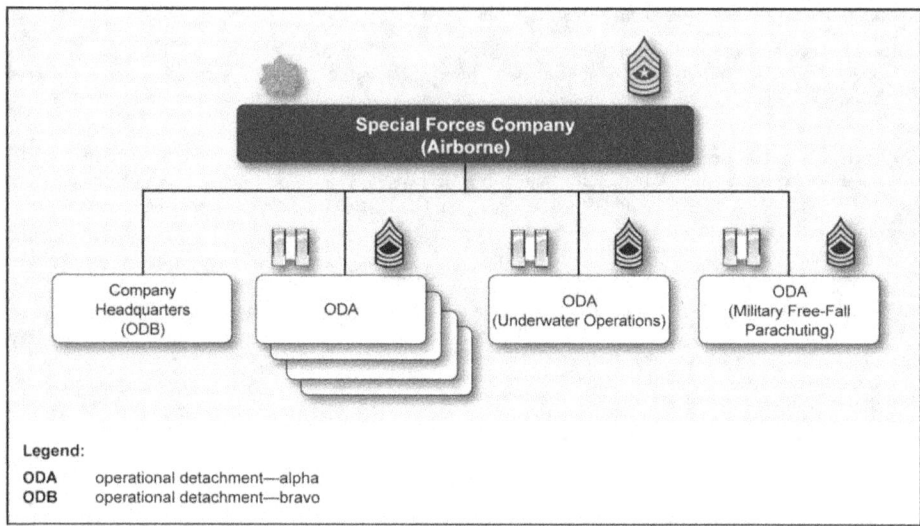

Figure 4-11. Special Forces company (airborne) organization

Organization

4-48. SF companies consist of an ODB and six ODAs. The ODB is a 15-man team. A typical SF company has one ODA trained in underwater operations and one ODA trained in military free-fall. The remaining ODAs may be trained in SF military mountaineering, surface maritime operations (water infiltration and scout swimmer), or mobility operations based on each ODA mission-essential task list.

> ## Operation SAFE HAVEN
>
> Operation SAFE HAVEN took place in the Republic of Panama from September 1994 to March 1995. The mission was to relieve the overcrowding of Cuban and Haitian migrant camps located at the U.S. Naval Base in Guantanamo Bay, Cuba. To accomplish this mission, the U.S. Southern Command established JTF-SAFE HAVEN on Empire Range in Panama.
>
> JTF-SAFE HAVEN began to build four camps to house and feed 10,000 Cuban migrants, with two camps to be commanded by U.S. Army personnel, one camp to be commanded by U.S. Air Force personnel, and one camp to be commanded by U.S. Navy personnel.
>
> Because of their language skills, cultural understanding, and their ability to operate in an ambiguous situation, SF Soldiers from C Company, 1st Battalion, 7th SFG, were deployed to Panama to operate in the migrant camps and interact directly with the Cuban migrants being flown to Panama from Guantanamo Bay.
>
> *(continued)*

> ## Operation SAFE HAVEN (continued)
>
> Once the JTF headquarters and camps were built, the ODB was collocated with and supported the JTF-SAFE HAVEN headquarters while the ODAs from the company were placed under the tactical control of the various Army, Navy, and Air Force camp commanders. The SF Soldiers worked alongside other SOF from the 96th CA Battalion and the 4th Psychological Operations Group inside of these camps.
>
> Although this was not a typical SF mission, the maturity and Spanish-speaking abilities of SF Soldiers enabled them to serve as liaisons between U.S. military personnel and Cuban residents, to defuse potential harmful situations before they came to friction, and to resolve numerous domestic problems and disturbances among the Cubans themselves.
>
> ODA personnel were responsible for the maintenance of living quarters for the Cuban migrants, sanitation, water usage, and the overall health and welfare of the Cuban population.
>
> When growing frustrations among the Cuban migrants led to disturbances and riots in the camps, the SF Soldiers were instrumental in calming the situation and restoring order to the camps. When Operation UPHOLD DEMOCRACY was successful in removing a military regime in Haiti, Haitian migrants were returned to Haiti relieving overcrowding at Guantanamo Bay. This allowed the Cuban migrants to be returned to Guantanamo Bay from Panama before their eventual immigration to the United States. By effectively interacting with the Cuban migrants and mediating peaceful solutions to many different problems, SF Soldiers were able to effectively enhance the humanitarian nature of the mission.

Function

4-49. In garrison, the ODB mission commands its own organic ODAs. When deployed, the ODB functions as a separate operational detachment with its own assigned mission. The mission may require the ODB to operate separately or to exercise operational control of a mix of organic and attached ODAs or other JSOTF assets. The SF company commander is an experienced SF major.

Characteristics

4-50. The SF company has the capacity to—
- Train and prepare ODAs for deployment.
- Infiltrate and exfiltrate an operational area by air, land, and sea.
- Conduct operations in remote areas for extended periods with minimal external direction and support.
- Develop, organize, equip, train, and advise or direct indigenous forces of up to regimental size.
- Serve as a SOCCE or SF liaison element at a functional component or Service component headquarters.
- Establish and operate an AOB to expand the command and control capabilities of the JSOTF or SOTF.
- Establish and operate an isolation facility for a JSOTF or SOTF.

4-51. Because the ODB (Figure 4-12, page 4-18) is relatively small, it may require augmentation to perform functions such as serving as a SOCCE or SF liaison element, or establishing and operating an AOB or isolation facility. The ODB is not designed to operate as a split element.

Chapter 4

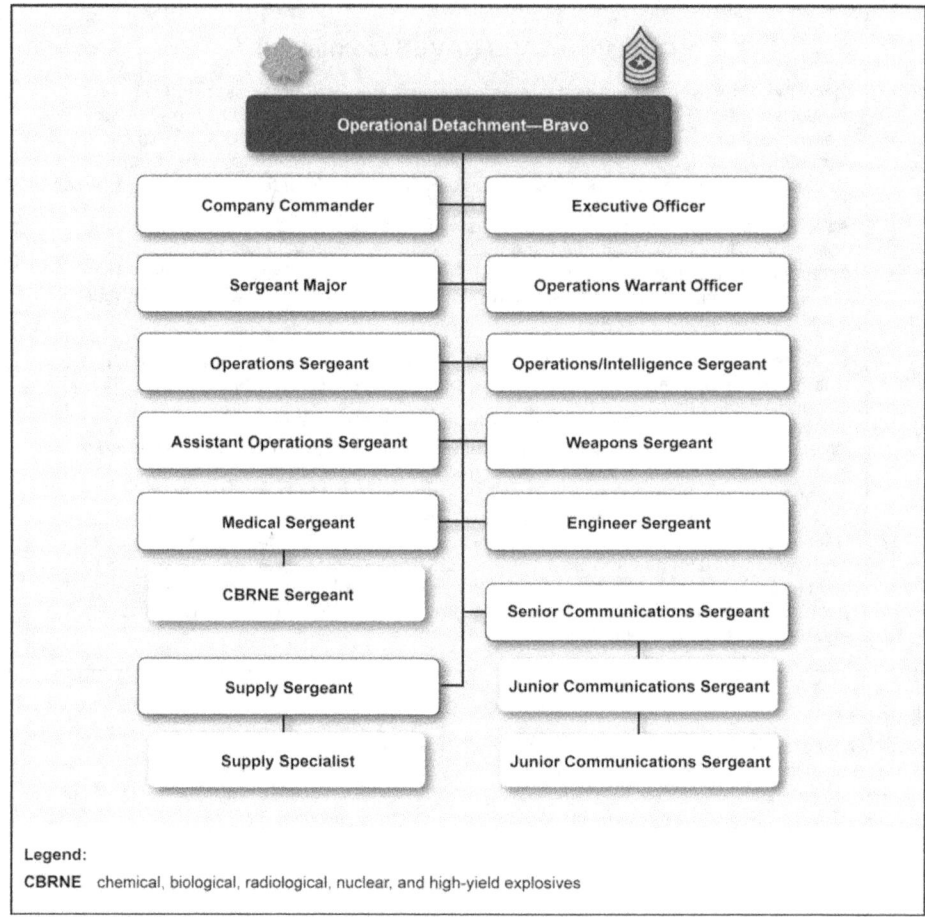

Figure 4-12. Special Forces operational detachment—bravo organization

Crisis Response Force

4-52. The crisis response force is a regionally oriented force required to respond to a crisis and conduct contingency operations within its designated regional joint operations area. Each GCC has one SF company dedicated for this purpose. The crisis response force is under operational control of its regional GCC and under administrative control of its own organic SFG.

4-53. All SF personnel assigned to a crisis response force have prior experience on ODAs and are graduates of the SF Advanced Reconnaissance, Target Analysis, and Exploitation Techniques Course. Select personnel within the crisis response force are also graduates of the SF Sniper Course. Detachment commanders are senior captains who have previous detachment command experience and are selected by both battalion and group commanders for these positions. Assistant detachment commanders are veteran SF warrant officers who have prior assistant detachment commander experience and prior experience in a crisis response force as a noncommissioned officer. These warrant officers are selected for these positions by their battalion and group senior warrant officers and battalion and group commanders. Detachment

Organization

operations sergeants are noncommissioned officers who have prior crisis response force experience. Crisis response force company commanders and company sergeants major are selected based on being SF Advanced Reconnaissance, Target Analysis, and Exploitation Techniques Course graduates and having previous crisis response force experience.

SPECIAL FORCES OPERATIONAL DETACHMENT—ALPHA

4-54. The ODA (also known as an ODA, A detachment, or A team) is composed of 12 men. It is the primary SF operational unit and the building block for SF operations. All other SF organizations are designed to command, control, and support the ODA.

Mission

4-55. The ODA is designed to organize, equip, train, advise or direct, and support indigenous military or paramilitary forces engaged in UW or FID activities. Using its inherent capabilities, the ODA also performs the other SF principal tasks discussed in Chapter 3. The time-tested composition of the ODA remains as viable and relevant today as it has been throughout the history of SF.

Organization

4-56. The commander of a 12-man ODA is a captain. Figure 4-13 shows the ODA organization. Other key leadership on an ODA are the assistant detachment commander (an SF warrant officer) and the operations sergeant (a master sergeant). The ODA has one intelligence sergeant and two specialists in each of the four primary SF functional areas—weapons, engineer, medical, and communications.

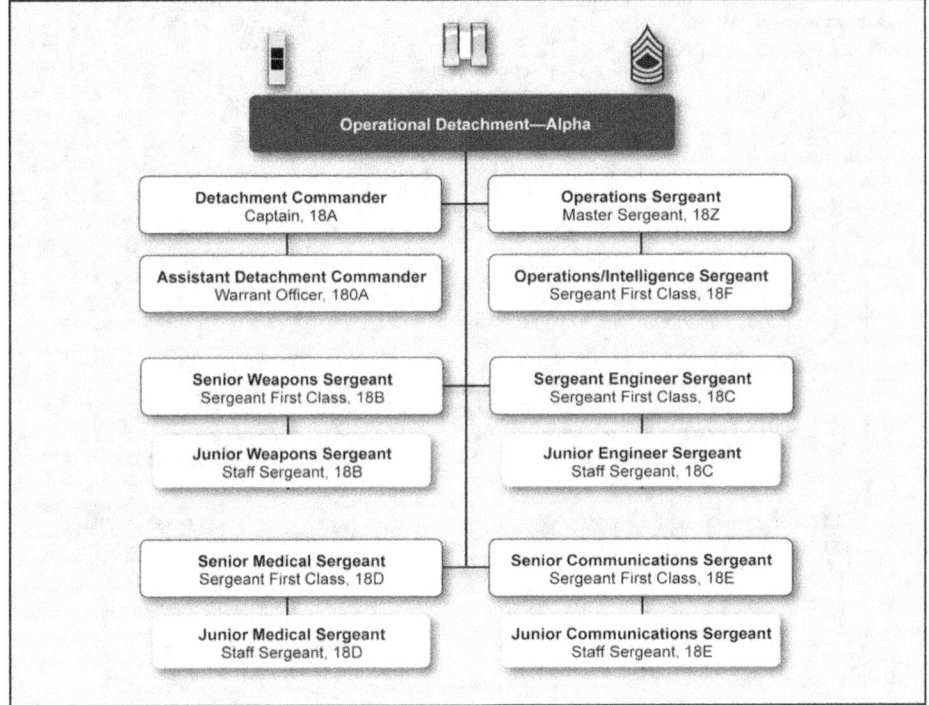

Figure 4-13. Special Forces operational detachment—alpha organization

4-57. Just as conventional units task organize to perform specific missions, SF commanders also task organize in response to mission requirements. SF commanders can tailor an ODA when an integral ODA is not the optimal organization to perform a mission or when a mission requires a mix of skills not found in an integral ODA. There are two means of tailoring ODAs: split-team operations and composite teams.

Split-Team Operations

4-58. Split-team operations allow the ODA to be employed in numerous configurations. The redundant capabilities of the four primary SF functional areas within an ODA allow the detachment commander to employ his detachment in a split-team configuration, where an ODA is divided into two or more operational elements, each capable of conducting sustained operations on a reduced scale. A detachment commander may opt for split-team operations when—

- The assigned mission or situation does not warrant the employment of the full team.
- The hostile situation does not permit operations by a complete ODA.
- A change in situation requires a deployed ODA to split and execute an additional mission.

4-59. The task organization for split-team operations are situational or mission dependent. In one example of a simple split-team operation, the detachment commander and the assistant detachment commander each command one of the split teams. The operations sergeant and the second senior-ranking noncommissioned officer on the detachment serve as the senior noncommissioned officers on the split teams.

Composite Teams

4-60. Certain SF missions require the organization of composite teams. An SF commander will usually reserve the right to tailor a composite team dependent upon the situation and mission. The SF commander will draw individual Soldiers from established ODAs to obtain the proper mix of skills and experience needed to conduct a specific mission. The SF commander must balance mission requirements against the effects on unit morale, readiness, and operations security before organizing composite teams. Additional SF, SOF, or other personnel (other government agencies, conventional forces, or combined forces) may augment ODAs to form composite teams.

Split-Team Operations

On the evening of 19 October 2001, the first of several SF elements infiltrated Afghanistan. That same night, the 12 men of ODA 595 infiltrated the Darya Suf Valley on MH-47s to join General Dostum's forces in Dehi, some 60 miles south of Mazar-e Sharif. Not long after, the team split into two sections—one accompanying Dostum to his headquarters, the other remaining at Dehi.

In late October 2001, General Franks, Commander, U.S. Central Command, met with the Northern Alliance commander Fahim Khan in Tajikistan and both had eventually agreed to focus operations in the north, specifically against the cities Mazar-e Sharif, Taloqan, and Konduz. By the time Franks and Fahim met, JSOTF-N had inserted two ODAs and a battalion headquarters element in the Mazar-e Sharif vicinity to help the Northern Alliance capture the city.

ODA 595, which had joined up with General Dostum south of the city, wasted little time calling in its first series of airstrikes on 21 October 2001 against Taliban positions in the Beshcam area, about 8 miles from Dostum's headquarters.

(continued)

Organization

> **Split-Team Operations (continued)**
>
> Captain Mark Nutsch, the ODA 595 team leader, then moved his men forward to Chobaki and directed additional airstrikes on Taliban tanks, artillery, and a command post near Chapchal. In quick succession, Northern Alliance forces took control of several villages in the district south of Mazar-e Sharif, assisted in great measure by additional airstrikes directed by ODA 595.
>
> On 26 October, Captain Nutsch sent a three-man element to Omitak Mountain to intercept enemy troops moving south toward Dostum's forces and to conduct further airstrikes. Then, during the evening of 28 October, a U.S. Air Force tactical air control party arrived, allowing Captain Nutsch to split ODA 595 into four three-man elements along with a two-man mission command cell composed of himself and a radio operator. The next day saw the arrival of a headquarters element from 3d Battalion, 5th SFG, commanded by Lieutenant Colonel Max Bowers, to provide mission command support for ODA 595, ODA 534, and General Dostum's Northern Alliance forces in preparation for the impending battle.
>
> During the week of 29 October, ODA 595 teams spread across the region south of Mazar-e Sharif to prepare for the final assault. On 5 November, General Dostum's men were ready to move. The operation began at dawn when MC-130 aircraft dropped two 15,000-pound BLU-82 "Daisy Cutter" bombs on Taliban locations at Aq Kuprok. However, one of the ODA 595 teams had crept close to Taliban positions and the Taliban commander counterattacked, attempting to trap the team. Close air support, Joint Direct Attack Munitions, and strafing runs by F-14s disrupted the Taliban attack and assisted the SOF in escaping safely. Other ODA 595 teams had similar success directing airstrikes from B-52s, a Predator unmanned aerial vehicle, and other aircraft against key Taliban command and control sites.
>
> On 9 November, General Dostum began his final push. In preparation for the assault, ODA 595 Soldiers called B-52 strikes on Taliban defenders who were dug in on the reverse slope of a ridge outside the city. But, by late afternoon, Northern Alliance forces fought off last-ditch Taliban counterattacks and, led by Captain Nutsch on horseback, seized the ridge. The next day, Northern Alliance troops seized the city airport allowing General Dostum and his SF advisors to ride into Mazar-e Sharif where they were greeted warmly by the population. This quick victory would have been highly unlikely without the marriage of Northern Alliance forces and coalition air power that the ODAs made possible. Even the Taliban regime tacitly acknowledged the role that airpower had played in the taking of Mazar-e Sharif. On 10 November, the Bakhtar News Agency quoted a Taliban official as stating, "For seven days continuously they have been bombing Taliban positions. They used very large bombs."
>
> Paraphrased from Dr. Donald P. Wright
> *A Different Kind of War: The United States Army in Operation ENDURING FREEDOM, October 2001–September 2005*
> Fort Leavenworth, Kansas: Combat Studies Institute Press, 2010

Function

4-61. ODAs can operate independently or with indigenous forces within a denied area. The ODA has many functions. Among these, the ODA can—
- Plan and conduct SF operations separately or as part of a larger force.
- Infiltrate and exfiltrate specified operational areas by air, land, and sea.

- Conduct operations in remote or denied areas for extended periods of time with a minimum of external direction and support.
- Develop, organize, equip, train, and advise or direct indigenous forces up to battalion size.
- Train, advise, and assist other U.S. and multinational forces and agencies.
- Plan and conduct unilateral SF operations.
- Perform other special operations activities as directed by higher authority.

Characteristics

4-62. Cross-training is fundamental to SF Soldiers. All team members conduct cross-training within the ODA in their various SF military occupational specialty skills. These additional skills are important in maintaining split-team operations and a mission-capable status in the event of a casualty.

4-63. To ensure functional coverage of all pre-mission activities, the ODA commander organizes a detachment staff using the mission variables—mission, enemy, terrain and weather, troops and support available, time available, and civil considerations. Detachment staff members perform their functional duties as described in ADRP 5-0. Because of the limited number of personnel on an ODA, they are of limited use as a strike, long-term, or permanent occupation force. Certain missions may require augmentation from a SOTF, JSOTF, or conventional forces.

SPECIAL FORCES SOLDIER

4-64. The SF Soldier has undergone a careful selection process prior to beginning his SF career. During the SF Assessment and Selection course, a future SF Soldier must demonstrate distinguishing core attributes, many that are derived from the UW mission. These attributes have evolved over the years because of changing mission requirements and focus by the GCCs to dictate the needs of SF training. These SF core attributes (described in Table 4-1, page 4-23) make SF the force of choice for complex, difficult, high-risk, and politically sensitive missions.

4-65. Every SF Soldier maintains a high degree of proficiency in cultural awareness, including a language capability, military occupational specialty skills, and advanced skills. Each SF Soldier is multifunctional and multicapable. Although trained as a specialist in a primary military occupational specialty, this Soldier is cross-trained in each of the SF specialties. Advanced skills are also taught within SF to enhance the operating capabilities of the force. Each SF unit conducts extensive area and country studies. From full combat operations, such as Operations ENDURING FREEDOM and IRAQI FREEDOM in Afghanistan and Iraq, respectively, through peacekeeping and peace enforcement in Bosnia and Haiti, to humanitarian assistance and disaster relief, such as Operation SAFE HAVEN in Panama and Operation SEA ANGEL in Bangladesh, SF teams are usually the first forces on the ground and the last to leave.

Cultural Awareness

4-66. SFGs are regionally aligned units that have global presence. SF Soldiers are educated and trained to develop and sustain long-term relationships with indigenous personnel and therefore create a cadre of language and culturally astute Soldiers who provide TSOC commanders, U.S. Ambassadors, and follow-on forces with critical capabilities and knowledge. Therefore, SF Soldiers possess certain cross-cultural communications traits that permeate the dialogue and the relationships they build. Through this cultural awareness, the SF Soldier facilitates national strategic objectives through face-to-face key leader engagements.

4-67. SF personnel are focused on developing and employing foreign forces and other assets in support of U.S. policy objectives. As a result, SF recognizes that gaining and maintaining cultural competency is critical. Cultural competence comprises the following three components:
- Awareness of one's own cultural worldview and attitude toward cultural differences.
- Knowledge of different cultural practices and worldviews.
- Cross-cultural skills.

Organization

Table 4-1. Special Forces core attributes

Attribute	Example
Integrity	Is trustworthy and honest; acts with honor and unwavering adherence to ethical standards.
Courage	Acts on own convictions despite consequences; is willing to sacrifice for a larger cause; is not paralyzed by fear of failure.
Perseverance	Works toward an end; has commitment; possesses physical or mental resolve; is motivated; gives effort to the cause; does not quit.
Personal Responsibility	Is self-motivated and an autonomous self-starter; anticipates tasks and acts accordingly; takes accountability for own actions.
Professionalism	Is a standard-bearer for the regiment; has a professional image, to include a level of maturity and judgment mixed with confidence and humility; forms sound opinions and makes own decisions; stands behind own sensible decisions based on own experiences.
Adaptability	Has the ability to maintain composure while responding to or adjusting own thinking and actions to fit a changing environment; possesses the ability to think and solve problems in unconventional ways; has the ability to recognize, understand, and navigate within multiple social networks; has the ability to proactively shape the environment or circumstances in anticipation of desired outcomes.
Team Player	Is able to work on a team for a greater purpose than self; is dependable and loyal; works selflessly with a sense of duty; respects others and recognizes diversity.
Capability	Has physical fitness, to include strength and agility; has operational knowledge; is able to plan and communicate effectively.

4-68. Developing cultural competence results in an ability to understand, communicate with, and effectively interact with people across cultures. For SF, this translates to building cultural adaptability as defined by Table 4-2, page 4-24.

4-69. SF cultural competency is measured in four modalities: interpersonal skills, nonverbal communication skills, language proficiency, and regional orientation. Each of these is discussed in the following paragraphs.

Interpersonal Skills

4-70. Interpersonal skills are critical to SF operations. They require the ability to listen with understanding, the ability to maintain an open mind, and the sensitivity to observe and grasp the essential components of a given situation. SF Soldiers combine the ability to overcome ethnocentricity and to treat indigenous forces as equals, while communicating and teaching across intercultural barriers. SF Soldiers use their interpersonal skills to obtain and maintain appropriate relationships with partner-nation counterparts. Obviously, interpersonal skills are difficult to define and to quantify; however, SF Soldiers possess the ability to interact with indigenous personnel through charismatic and personal engagement.

Political Awareness

4-71. Sometimes referred to as "warrior diplomats," SF Soldiers maintain a keen appreciation of the political aspects of their operational environment. They must understand U.S. policies, goals, and objectives and be able to articulate them in a manner that convinces their HN counterparts to support them. Similarly, they must understand the political context within which their counterparts operate. Politico-military considerations will frequently shape SF operations, requiring clandestine, covert, or low-visibility techniques, and oversight at the national level.

Table 4-2. Adaptability competencies

Competency	Example
Physically Oriented Adaptability	• *Environmental resolve*: Adjusting to environmental states. • *Physical resolve*: Pushing oneself to complete physical tasks. • *Physical fitness*: Adjusting weight/strength, as necessary.
Learning Tasks	• *Proactively*: Searching out learning opportunities in advance of a challenge. • *Reactively*: Doing what is necessary to keep up. • *Self-Reflection*: Taking action to improve through the process of reflection.
Handling Emergencies or Crises	• *Responding to life-threatening emergencies*: For example, a sniper. • *Administering first aid*: Applying medical techniques.
Handling Stress	• *Resilience and emotional control*: Remaining calm and in control of emotions. • *Extraordinary stressors*: Reacting under fire, extreme noise, negative feedback, and so on.
Dealing With Change or Ambiguity	• *Cognitive flexibility*: Responding to changing environments; adjusting one's perspective. • *Flexible decisionmaking*: Changing approach to a problem. • *Critical thinking*: Identifying powerbrokers, motives, and leverage points; displaying situational awareness and understanding.
Thinking Creatively	• *Problem-solving*: Developing innovative solutions. • *Working with limited resources*: Developing methods of obtaining or using resources. • *Information gathering*: Using innovative methods to understand links and capitalize on opportunities.
Cultural Adaptability	• *Negotiation of cultural/language barrier*: Communicating effectively across cultures. • *Cultural tolerance*: Displaying a willingness to adjust behavior or show respect for others' customs or values.
Interpersonal Adaptability	• *Building rapport*. • *Influencing the dilemma/negotiation*: Communicating with awareness of leverages. • *Membership and participation as a team member*: Working with others.

Problem Solving

4-72. The nature of UW and other SF missions often defies template or "schoolhouse" solutions. A hallmark of the SF Soldier is the ability to analyze a situation, then adapt and apply U.S. doctrine, tactics, techniques, procedures, equipment, and methods in a culturally sensitive and appropriate manner to resolve difficult issues in nonstandard situations.

Nonverbal Communication Skills

4-73. Nonverbal communications are wordless messages—the way people communicate by sending and receiving signals (for example, body language, eye movements, gestures, postures, proximity, facial expressions, and symbols). Similar to verbal communications, nonverbal communications differ across cultures, although some signals are shared. The impact of a communication message can be broken down into 7 percent verbal (words), 38 percent vocal (volume, pitch, rhythm, and so on), and 55 percent body movements (mostly facial expressions). SF Soldiers learn to communicate nonverbally without acquiring a foreign language through a good understanding of a population's nonverbal communication systems.

Language Proficiency

4-74. Language proficiency is a key component in cross-cultural communications. Each prospective SF Soldier is tested for language ability through either the Defense Language Proficiency Test or the Oral Proficiency Interview. The Defense Language Proficiency Test measures reading and listening skills and the Oral Proficiency Interview measures participatory and active conversation. To graduate from the SF Qualification Course, each prospective SF Soldier must attend language curriculum and/or pass one of these measurement tests. The desired portrait of an SF Soldier is one who can gain and maintain rapport with indigenous personnel; learning a language through participatory and active conversation remains a priority over reading and listening. Though participatory dialogue remains a priority, it is not wise to discount the fact that an SF Soldier may be required to translate documents or listen to intercepted conversations. This education is merely a stepping stone as SF Soldiers continue to improve their language skills through routine and dedicated unit-sponsored training, immersion training, individual study, and repeated deployments to their region of orientation. As a result, all SF Soldiers possess varying levels of language ability in one or more foreign languages.

Regional Orientation

4-75. SF units are regionally oriented to ensure they have the resident skills and knowledge of the belief, art, morals, law, custom, and any other capabilities and habits of a specific region to allow them to influence their HN counterparts. This understanding of the region extends into the political, military, economic, social, infrastructure, information, and physical environment systems within that region and how these systems affect military operations. Formal training and cultural immersion during repeated deployments are the vehicles for developing this understanding.

Members

4-76. The following paragraphs describe the members of the ODA and outline their key responsibilities within the team.

Detachment Commander

4-77. The ODA commander is an Army captain with the military occupational specialty of 18A; the commander has the overall responsibility for mission success or failure. Responsible for the planning and the execution of the mission, the commander ensures that the ODA mission and his intent are nested (two levels up). Often the senior representative of U.S. interests in foreign countries, the commander is an expert in all things related to UW and counterinsurgency operations. The commander is an adept planner and tactician and is responsible for advising or commanding combat forces up to a battalion-sized element.

Assistant Detachment Commander

4-78. The assistant detachment commander is an SF warrant officer, military occupational specialty 180A. The assistant detachment commander serves as the second in command or commands in the absence of the detachment commander during split-team operations and may command during composite team operations. The assistant detachment commander position is unique to SF and, as the focal point for synchronization of detachment staff functions, has responsibilities similar to that of a chief of staff, although these roles are not identical. The assistant detachment commander provides technical, tactical, and operational expertise; provides advice and assistance to the detachment commander and detachment personnel across the special operations continuum; and is an experienced subject-matter expert in UW. Primary responsibilities within the detachment center around directing operational and intelligence fusion, the planning process for current and future detachment engagements, area studies and intelligence collection efforts, mid- and long-range training management, personnel recovery, integration of new technologies, interagency and intergovernmental relationships, and the application of advanced special operations. The assistant detachment commander also is capable of advising or commanding, directing, or leading indigenous forces up to battalion size.

Operations Sergeant

4-79. The operations sergeant, military occupational specialty 18Z, is the senior enlisted member of the detachment. Similar to the staff roles and responsibilities of the operations officer, the operations sergeant is responsible for the day-to-day activities of the detachment. Responsibilities of the operations sergeant include advising the detachment commander on all operations and training matters and providing leadership, tactical and technical guidance, and professional support to detachment members. With guidance from the commander, the operations sergeant assigns specific tasks, supervises the performance of detachment tasks, and prepares plans, orders, and reports. The operations sergeant oversees individual and collective training and prepares the operations and training portions of area studies, briefbacks, operation plans, concept of operations, and operation orders, and supervises the detachment in the preparation of these documents. The operations sergeant writes daily training schedules and maintains responsibility for short-term training, and can organize, train, assist, advise, or lead indigenous forces up to battalion size.

Assistant Operations and Intelligence Sergeant

4-80. The assistant operations and intelligence sergeant, military occupational specialty 18F, has responsibilities similar to that of the intelligence officer, and is the detachment member responsible for all aspects of intelligence, counterintelligence, and protection for the ODA and its indigenous forces. Assistant operations and intelligence sergeants will plan, coordinate, and conduct continuous collection planning and intelligence analysis in support of the detachment's area study and intelligence files effort. During mission planning, the assistant operations and intelligence sergeant analyzes the detachment mission and evaluates the unit intelligence database and target intelligence package. The intent of this analysis and evaluation is to identify information and intelligence gaps and to conduct a detailed operational preparation of the environment in support of the development of the plan of execution. The assistant operations and intelligence sergeant requests imagery, maps, weather information, topographic terrain analysis products, and intelligence updates from the battalion intelligence staff officer, and prepares the evasion plan of action, disseminates the mission classification guidance, and assists the commander in implementing operations security and information security procedures.

4-81. While deployed, the assistant operations and intelligence sergeant continually updates the mission intelligence estimate and advises the detachment commander on significant changes in the threat and the HN military and civilian populace. The assistant operations and intelligence sergeant also tactically questions and processes enemy prisoners of war and civilian detainees, debriefs friendly patrols, conducts informal intelligence liaison with U.S. intelligence agencies and local HN military and police forces, and questions the local populace to acquire combat and protection information. The assistant operations and intelligence sergeant provides intelligence reports and summaries to higher headquarters, to include assisting in preparing and completing the Special Operations Debriefing and Retrieval System. As the subject-matter expert in biometrics, the assistant operations and intelligence sergeant and is responsible for collecting, processing, and analyzing biometric information. In addition, the assistant operations and intelligence sergeant advises the commander concerning the use of MIS and CA in support of the information operations campaign. The assistant operations and intelligence sergeant assists the detachment operations sergeant in preparing area studies, briefbacks, operation plans, concept of operations, and operations orders, and may assume the duties of the operations sergeant during split-team operations. The assistant operations and intelligence sergeant can organize, train, assist, direct, or lead indigenous forces up to company size.

Weapons Sergeant

4-82. The two weapons sergeants, military occupational specialty 18B, are tactical mission leaders that know tactical training and range fire as well as how to teach marksmanship and the employment of weapons systems to others, employing tactics and techniques in all aspects of conventional and UW environments. They have similar staff roles and responsibilities as those of the assistant operations staff officer.

4-83. When deployed, the weapons sergeants train detachment members and indigenous forces in light infantry tactics and in the use of small arms, submachine guns, machine guns, grenade launchers, forward observer procedures, indirect-fire weapons such as mortars, and anti-armor weapons systems found throughout the world. They assist the detachment operations sergeant in planning and implementing the tactical security plan of the detachment, and they aid the detachment operations sergeant in preparing area studies, briefbacks, operation plans, concept of operations, and operations orders. Weapons sergeants can organize, train, assist, direct, or lead indigenous forces up to company size.

Organization

Engineer Sergeant

4-84. The two engineer sergeants, military occupational specialty 18C, perform, lead, supervise, instruct, advise, and maintain proficiency in all tasks in demolitions, explosives, improvised munitions, unexploded ordnance, construction, field fortification, rigging, electrical wiring, reconnaissance, and target analysis in all aspects of conventional and UW environments. They serve as the logistics staff noncommissioned officers and are responsible for all logistics matters on an ODA. The engineer sergeants prepare and review target analysis folders and are responsible for the planning, execution, and supervision of cross-training of detachment members in SF engineering skills. They supervise combat engineering functions when conducting split-team operations and missions.

4-85. When deployed, because of homemade explosives training, engineer sergeants are able to construct and employ improvised munitions. The engineer sergeants provide tactical and technical guidance to the detachment commander on combat engineer capabilities and functions, as well as train, advise, and supervise indigenous and multinational personnel. They will plan, instruct, and perform sabotage operations with standard, nonstandard, and improvised munitions and explosives. They plan, organize, train, advise, assist, and supervise indigenous and multinational personnel on engineering tactics on defensive and offensive techniques. The engineer sergeants plan, prepare, and conduct the target portion of the area study, operation plans, concept of operations, operations orders, and also conduct briefings, briefbacks, and debriefings. They can organize, train, assist, direct, or lead indigenous forces up to company size.

Medical Sergeant

4-86. The two SF medical sergeants, military occupational specialty 18D, provide emergency, routine, and limited definitive care for detachment members and associated indigenous personnel, and serve as the personnel staff officer. They train, advise, and direct detachment routine, emergency, and preventative medical care. They establish field medical facilities to support detachment operations. SF medical sergeants are highly trained as practitioners in emergency and primary care medicine, advanced life support techniques, parisitology, dentistry, and have the ability to deliver limited veterinary care. The primary limitation of SF medical sergeants is that they cannot open the thoracic, abdominal, or cranial cavities to control bleeding without the immediate availability of higher-level surgical support.

4-87. When deployed, medical sergeants serve as the medical officer for an indigenous force up to battalion size. Their duties within the joint special operations area can include, but are not limited to, overseeing pre-induction screening, combat lifesaver/tactical combat casualty care and other medical training, Soldier and family member healthcare, combat health support, and field preventative medicine. Indigenous medical personnel in the joint special operations area can augment the SF medical sergeant's efforts. The medical sergeants prepare the medical portion of area studies, briefbacks, operation plans, concept of operations, and operations orders, and can organize, train, assist, direct, or lead indigenous forces up to company size.

Communications Sergeant

4-88. The two SF communications sergeants, military occupational specialty 18E, install, operate, supervise, maintain, and control all communications and electronics equipment. They must have a thorough grounding in communication basics, communication procedures, computer technology, assembly and systems applications, antenna theory, and radio wave propagation. Their staff roles and responsibilities are similar to those of a conventional unit's signal officer as they will advise the commander on all communications-related matters during both mission planning and execution. They must understand communication theory and maintain proficiency in special operations communications techniques and equipment, to include how to install, operate, and maintain frequency modulation, amplitude modulation, high frequency, very high frequency, and ultrahigh frequency radio systems, the Joint Base Station, and new computer technology.

4-89. When deployed, communications sergeants must be able to make communications in voice and data, continuous wave, and burst radio nets by utilizing computer systems and networks. They are experts in sending and receiving critical messages that link the ODA with its mission command elements. They will train detachment members and indigenous forces on communications equipment and procedures, and will prepare the communications portions of SF area studies, briefbacks, operation plans, concept of operations, and operations orders. They can organize, train, assist, direct, or lead indigenous forces up to company size.

Military Education and Military Training

4-90. The individual SF Soldier is both highly educated and highly trained. Military education is the instruction of individuals in subjects that will enhance their knowledge of the science and art of war, and the profession of arms. Military training, on the other hand, is the preparation of individuals or units to enhance their capacity to perform specific military functions, tasks, or missions. In other words, the education provided to an SF Soldier will provide the individual with the ability to determine *what* to do in a given circumstance while training will provide the *how*.

4-91. SF Soldiers require a combination and fine balance between both education and specialized training to achieve operational proficiency. Education enhances the understanding of SF operations and fosters a disciplined thought process and creativity needed to develop solutions to complex problems in dynamic and ambiguous situations. Training is designed to be both challenging and realistic in order to produce SF personnel and units that have mastered the tactics, techniques, and procedures through which SF units accomplish their missions.

4-92. Specialty skill training provides SF the advanced skills necessary to perform operational and force structure requirements. The SFGs conduct proficiency and sustainment training for selected advanced skills, such as military free-fall, SF Sniper Course, combat dive qualification, SF advanced urban combat, and operational preparation of the environment. Table 4-3, page 4-29, shows the advanced skills normally found on an ODA team.

Military Free-Fall

4-93. Military free-fall parachuting allows SF personnel to deploy their parachutes at a predetermined altitude, assemble in the air, navigate under canopy, and land safely together as a tactical unit ready to execute their mission. Military free-fall is primarily a means of entering a designated area within the objective area. There are two basic types of military free-fall operations:

- High-altitude/low-opening operations are jumps made with an exit altitude of up to 35,000 feet above mean sea level and a parachute deployment altitude at or below 6,000 feet above ground level. High-altitude/low-opening infiltrations are the preferred military free-fall method of infiltration when the enemy air defense posture is not a viable threat to the infiltration platform. High-altitude/low-opening infiltrations require the infiltration platform to fly within several kilometers of the drop zone.
- High-altitude/high-opening operations are standoff infiltration jumps made with an exit altitude of up to 35,000 feet above mean sea level and a parachute deployment altitude at or above 6,000 feet above ground level. High-altitude/high-opening infiltrations are the preferred method of infiltration when the enemy air defense threat is viable or when a low-signature infiltration is required. Standoff high-altitude/high-opening infiltrations provide commanders a means to drop military free-fall parachutists outside the air defense umbrella, where they can navigate undetected under canopy to the drop zone or objective area.

Advanced Special Operations Techniques

4-94. Advanced special operations techniques is an unclassified umbrella term used to describe all SOF operations that require the use of specific, classified tactics, techniques, and procedures.

Organization

Table 4-3. Advanced skills on a Special Forces operational detachment—alpha

Advanced Skill	Number per Special Forces Operational Detachment—Alpha
Language Qualification	All
Survival, Evasion, Resistance, and Escape Level C	All
Static-Line Jumpmaster	3 Each
Military Free-Fall Jumpmaster (military free-fall detachments A only)	3 Each
Diving Supervisor (underwater operations detachments only)	2 Each
Diving Medical Technician (underwater operations detachments only)	2 Each
Special Forces Sniper Course (Level I)	1 Each
Special Forces Sniper Course (Level II)	1 Each
Advanced Special Operations Techniques (Level II)	All
Advanced Special Operations Techniques (Level III)	2 Each
Special Technical Electronic Equipment	2 Each
Special Forces Advanced Urban Combat	All
Joint Terminal Attack Controller	1 Each
Joint Fires Observer	2 Each
Small Unmanned Aircraft System	2 Each
Special Operations Forces Site Exploitation (Operator Basic)	All
Special Operations Forces Site Exploitation (Operator Advanced)	1 Each
Mountaineering (Level I) (mountain detachments only)	1 Each
Mountaineering (Level II) (mountain detachments only)	3 Each
Mountaineering (Level III) (mountain detachments only)	8 Each

Waterborne Operations

4-95. Waterborne operations involve the employment of forces from air, land, and waterborne platforms to meet objectives ashore; they may be in support of or independent from amphibious operations. Waterborne operations are one of many options available to a commander to infiltrate/exfiltrate a detachment into/out of a designated area of operations for the purposes of executing any SF mission. ODAs are proficient in basic waterborne infiltration/exfiltration techniques, to include small-boat operations, surface-swimming operations, helocasting operations, riverine operations, and air operations.

4-96. Specific ODAs conduct waterborne operations along coastlines, coastal river junctions, harbors, or other inland waterways in order to exploit environmental characteristics found within an area of operations, and use these characteristics to their advantage. SF combat diver detachments remain proficient in all aspects of SF maritime operations. Maritime operations include basic and advanced open-circuit self-contained breathing apparatus operations, basic and advanced closed-circuit self-contained breathing apparatus operations, submarine operations, basic and advanced waterborne infiltration/exfiltration techniques, and a capacity to combine both air and diving platforms to accomplish a specific mission.

Advanced Mountaineering

4-97. Special operations advanced mountaineering is defined as an operation that requires an element to utilize specialized mountaineering techniques and equipment to accomplish a mission. USASOC Regulation 350-12, *Special Operations Forces Mountaineering Operations*, governs the conduct of mountaineering operations by all units assigned or attached to USASOC. Mountaineering consists of a multitude of subjective and objective hazards of which an SF Soldier must be continually aware while operating in a

Chapter 4

mountainous or alpine environment, and can provide a commander with a unique capability during peacetime or in wartime to obtain an objective goal. SFGs have ODAs specially trained to conduct advanced mountaineering operations, to include high-angle rescue operations. There are three qualification levels for the SF Soldiers assigned to these ODAs: Basic Mountaineer (Level III), Senior Mountaineer (Level II), and Master Mountaineer (Level I). These ODAs are trained to use a wide variety of skills and technical abilities, as well as specialized equipment and experience, in order to move a unit through mountain or alpine environments both effectively and safely.

Joint Terminal Attack Controller

4-98. Joint Terminal Attack Controllers are defined as qualified (certified) Servicemembers who, from a forward position, direct the action of combat aircraft engaged in close air support and other offensive air operations (JP 1-02). A Joint Terminal Attack Controller is the forward element of the theater air-ground system and must be organized, trained, and equipped to operate within that infrastructure.

4-99. A Joint Terminal Attack Controller is recognized across the DOD and other coalition nations as capable and authorized to perform terminal attack control. A Joint Terminal Attack Controller-certified and qualified SF Soldier or SF unit fire support noncommissioned officer has the capability to—

- Plan for close air support operations by having knowledge and understanding of close air support assets, capabilities, limitations, and employment.
- Prepare for close air support operations by being able to prepare equipment and applying products of operational planning to support close air support execution.
- Execute close air support operations, to include target acquisition, coordination, deconfliction, target marking, control procedures, and battle damage assessment.

Joint Fires Observer

4-100. Joint fires observers provide SF with the capability to exploit those opportunities that exist in the operational environment where a trained observer can efficiently support air-delivered fires and surface-to-surface fires and facilitate targeting. A joint fires observer-certified and qualified SF Soldier or SF unit fire support noncommissioned officer provides the SF commander the warfighting capability to—

- Request, control, and adjust surface-to-surface fires.
- Provide timely and accurate close air support targeting information to a Joint Terminal Attack Controller/forward air controller, or directly to aircraft when authorized by the controlling Joint Terminal Attack Controller/forward air controller.
- Perform and conduct autonomous terminal guidance operations as defined in JP 3-09.

4-101. The joint fires observer does not circumvent or nullify the need for a qualified Joint Terminal Attack Controller or forward air controller during close air support operations. To facilitate close air support execution, the joint fires observer should be regarded as an extension of the tactical air control party. In all cases, a qualified Joint Terminal Attack Controller/forward air controller will retain terminal attack control authority. Joint fires observers, in conjunction with Joint Terminal Attack Controllers, assist SF commanders with the timely planning, synchronization, and responsive execution of all joint fires and effects.

Small Unmanned Aircraft System

4-102. The operator of the small unmanned aircraft system meets all qualification standards and is mission-specific qualified to control a small unmanned aircraft system. The operator is trained in—

- General aerodynamics.
- Federal Aviation Administration/airspace/airfield operations.
- Visual flight rules.
- Airspace coordination and deconfliction.
- Safety of flight.
- Emergency procedures.

Organization

- Basic mission planning.
- Map reading and navigation.
- Weather.
- System overview.
- Hands-on flight and payload operations.
- Maintenance.

Special Operations Forces Site Exploitation

4-103. Personnel trained in the technical skills and operational procedures necessary to perform site exploitation in support of SOF missions are trained in site exploitation; biometrics; document and media exploitation; tactical questioning; CBRNE exploitation and awareness; cellular exploitation; and weapons technical intelligence. These personnel are cable of conducting site exploitation in both rural and urban environments.

4-104. USASFC recognizes four distinct categories of site exploitation execution, and each category has a different set of capabilities and associated equipment. The categories represent a tiered approach to site exploitation, each building on the capabilities of the tier below through additional training and advanced equipment. The first three categories are oriented toward on-target site exploitation collection and the fourth focuses on fixed-site exploitation and analysis. The four categories are—

- **Operator Basic (Level I).** Training at this level provides a working knowledge of assigned equipment tactics, techniques, and procedures to prevent site corruption and prepares the Soldier to properly document, exploit, and package the objective or site under SF control.
- **Operator Advanced (Level II).** Training at this level provides an enhanced capability to perform a detailed site exploitation with organic equipment. Qualified individuals possesses sufficient knowledge of SOF site exploitation to plan and run SOF site exploitation operations, to include performing or directing detailed searches, documentation, exploitation, and packaging of an SF-controlled site.
- **Enabler (Level III).** Training at this level results in an independent element that can be modified and attached to operational elements based on mission requirements. It maintains the same capabilities to plan, run, and instruct SOF site exploitation operations, but has advanced training and additional equipment permitting a more thorough site exploitation in a tactical environment.
- **Exploitation Analysis Center (Level IV).** Training at this level provides a capability to perform analysis and exploitation with advanced equipment in a laboratory environment. The personnel and equipment are organic to the SFG and perform as the core to an existing/standing exploitation analysis center. The exploitation analysis center mans and provides leadership elements for personnel assigned to the exploitation analysis center and performs detailed analysis with organic equipment or personnel, equipment, and data received from the tactical elements. The exploitation analysis center also provides access to interagency reachback analysis. Reachback capability is supported through the USSOCOM information technology architecture. The enabler element and exploitation analysis center is manned and attended by SFG chemical reconnaissance detachment and chemical decontamination detachment personnel.

Special Forces Advanced Reconnaissance, Target Analysis, and Exploitation Techniques Course

4-105. SF Advanced Reconnaissance, Target Analysis, and Exploitation Techniques Course is a classified nonsolicitation course available to Soldiers in career management field 18, warrant officers with military occupational specialty 180A, or officers with military occupational specialty 18A. SF Soldiers who are allowed to attend this course must have previously been assigned to an ODA for a minimum of 10 months, have a minimum clearance of Secret, and either be assigned or on orders to a crisis response force unit.

Chapter 4

Special Forces Sniper Course

4-106. The SF Sniper Course is designed to train individual SOF Soldiers in the technical skills and operational procedures necessary to deliver precision rifle fire from concealed positions to selected targets in support of SOF missions. Personnel are trained in advanced rifle marksmanship, sniper marksmanship, observation techniques, range estimation, advanced concealment methods, stalking, camouflage, target selection and interdiction, and integration in a direct action mission force. All subjects are taught in both rural and urban environments.

Special Forces Advanced Urban Combat

4-107. SF advanced urban combat provides precision combative techniques used on urban or complex terrain. These techniques include sustained skills in SF explosive breaching, selective and discriminatory target engagement, and emergency assault methods utilizing special nonstandard equipment and SF tactics, techniques, and procedures. It is characterized by offensive operations where combat in built-up areas is necessary, but minimal collateral damage is a primary consideration and noncombatants are or may be present. Operations are characterized by SF, special purpose weapons and equipment, a special personnel selection process, and advanced marksmanship.

Chapter 5
Employment

We continue to maintain that Special Forces Operational Detachments have the mission and capability of developing indigenous guerrilla forces, conducting operations behind the enemy lines, and of sustaining these operations for an indefinitely long time...the Chief of Staff has insisted that Special Forces shall not duplicate the training and doctrine of ranger and commando units.

Colonel Robert McClure

SF operates under many varied command relationships. The requirements at each echelon of command determine the exact structure. SF operations are inherently joint and frequently controlled by higher echelons, often with minimal involvement of intermediate headquarters. Certain functions or activities may require oversight at the national level. This chapter discusses employment and mission command of SF operations from the theater level and below. Mission command is "the exercise of authority and direction by the commander using mission orders to enable disciplined initiative within the commander's intent to empower agile and adaptive leaders in the conduct of unified land operations" (ADP 3-0). It is commander-led and blends the art of command and the science of control to integrate the warfighting functions to accomplish the mission. It provides a number of generic command relationships that may be tailored to meet the needs of the situation.

All SF commanders must understand the nature of joint, multinational, and interagency operations because their abbreviated chains of command often require frequent contact with joint and multinational headquarters as well as interagency departments. JPs 3-0 and 3-05 include detailed discussions of joint operations as they apply to SF operations. In order for SF to achieve unity of effort, coordination is required among nongovernmental organizations and intergovernmental organizations, and among multinational or coalition nations. An integral part of where all coordination begins is at the country team of the designated country.

COUNTRY TEAM

5-1. JP 1-02 defines country team as the senior, in-country, U.S. coordinating and supervising body, headed by the chief of the U.S. diplomatic mission (usually the U.S. Ambassador), and composed of the senior member of each represented U.S. department or agency, as desired by the chief of the U.S. diplomatic mission. Figure 5-1, page 5-2, shows the country team concept.

5-2. Members of the country team meet regularly to coordinate U.S. Government political, economic, and military activities in the HN. Many SF activities will cross the jurisdictional boundaries or responsibilities of other country team members. For this reason, an ODA will generally meet with members of the country team upon entry into any particular country to ensure that it is aligned and coordinated with the team's objectives and requirements.

Chapter 5

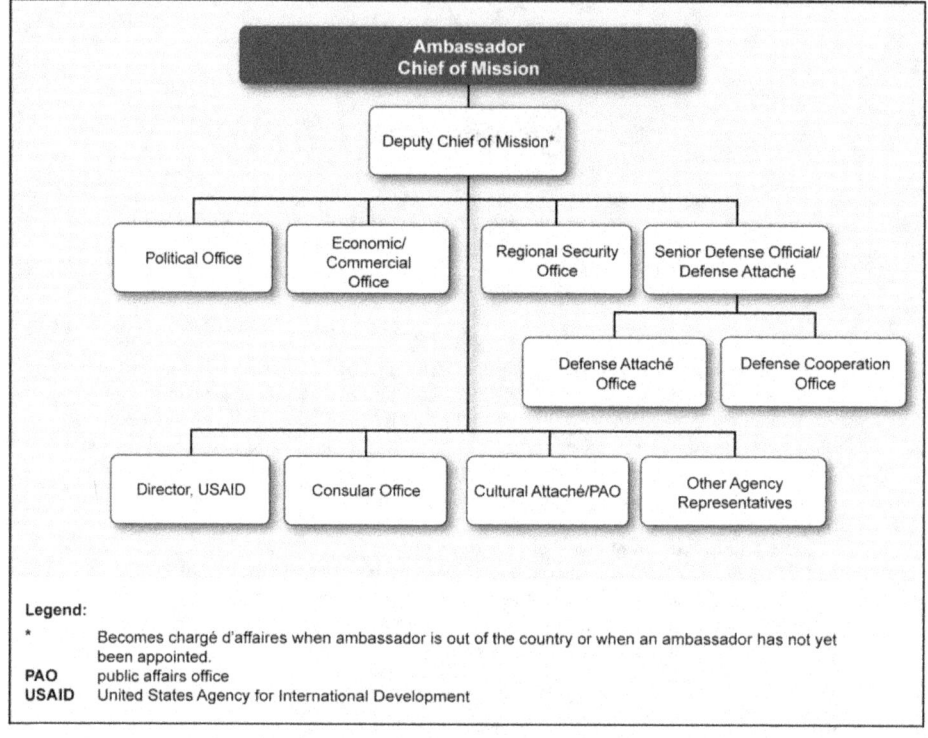

Figure 5-1. Country team concept

5-3. The United States maintains diplomatic relations with more than 180 foreign countries through embassies, consulates, and other diplomatic missions. The U.S. Ambassador to a country is responsible to the President for directing, coordinating, and supervising official U.S. Government activities and personnel in that country. These personnel include all U.S. military personnel not assigned to the combatant commander or other designated U.S. military area commander. Protection and security of U.S. military personnel are a matter of significant interest. Often, specific agreements are required between the U.S. Ambassador (also known as the chief of mission) and the GCC. ARSOF deployed to a particular country for various missions (exercise, operation, or security assistance) remain under the combatant command or under operational control (attached forces) of the GCC exercised through a subordinate headquarters (normally the TSOC). (JP 3-22 provides additional information.) Under no circumstances will SOF operate in a GCC's area of responsibility or in the U.S. Ambassador's country of assignment without prior notification and approval.

5-4. The GCCs are directly responsible to the President or the Secretary of Defense for the execution of assigned missions. The National Security Strategy and National Military Strategy, shaped by and oriented on national security policies, provide strategic direction for GCCs. In turn, GCCs plan and conduct unified actions in accordance with this guidance and direction (Figure 5-2, page 5-3). They ensure their joint operations are synchronized with other military forces (multinational operations) and nonmilitary organizations. Part of that synchronization can and, in most cases, will include special operations. SOF may be assigned to a GCC; however, command and control of special operations normally should be executed within a special operations chain of command.

Employment

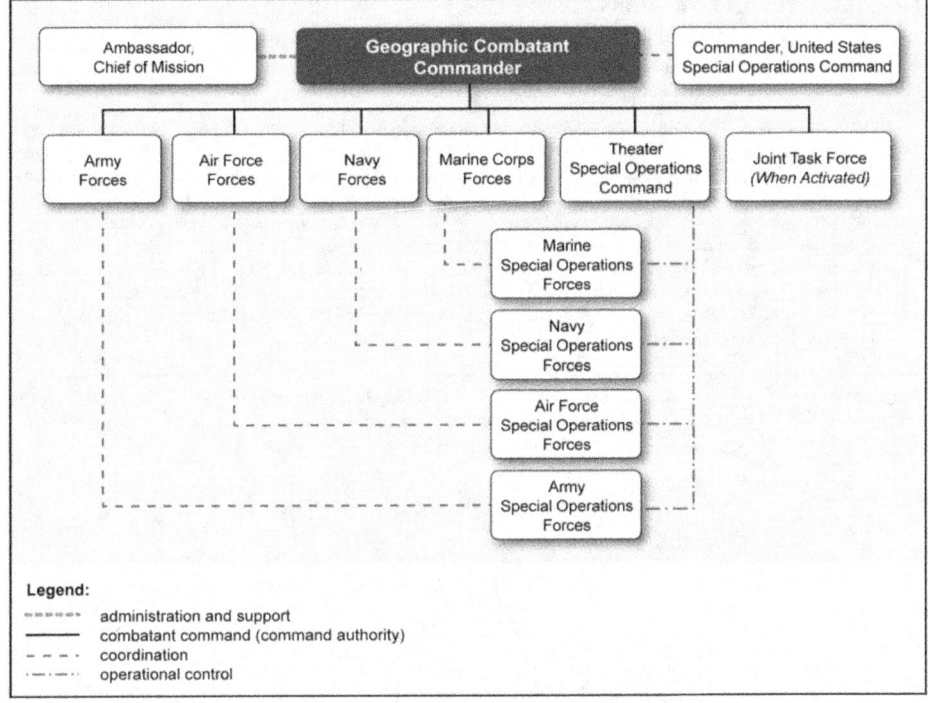

Figure 5-2. Theater command structure

5-5. The Secretary of State is the President's principal foreign policy advisor. In the National Security Council interagency process, the Department of State is the lead agency for most U.S. Government activities abroad. For this reason, the Department of State plays a key role in special operations.

5-6. Requests for SF or any other ARSOF may originate with the U.S. Ambassador, defense attaché, or security assistance organization chief from the country team who passes the requests through the appropriate GCC to the Chairman of the Joint Chiefs of Staff. The Chairman of the Joint Chiefs of Staff ensures proper interagency coordination. If the forces are available from joint operations area forces and no restrictions exist on their employment, the GCC can approve and support the request. If SF or SOF in general are insufficient in the joint operations area, the GCC can request the forces through the Joint Chiefs of Staff to the USSOCOM.

THEATER OF OPERATIONS ORGANIZATION

5-7. When the President or the Secretary of Defense authorizes military operations, the GCC organizes the area of responsibility to orchestrate joint operations with multinational and interagency activities. An integral part of this organization is the special operations staff element.

5-8. The interaction of the special operations area of responsibility staff element with ARSOF differs in each theater of operations because each GCC chooses to organize forces differently to meet the requirements of the specific strategic environment. Regardless of these organizational differences, the special operations area of responsibility staff elements all work closely with their TSOC in planning, directing, and conducting special operations missions and in integrating special operations into the area of responsibility strategy and campaign plan.

Chapter 5

THEATER SPECIAL OPERATIONS COMMAND

5-9. Normally, mission command of SOF is executed within the special operations chain of command. The identification of an organizational command structure for SOF should depend upon specific objectives, security requirements, and the operational environment.

5-10. The TSOC is the joint special operations command through which the GCC normally exercises operational control of SOF within the area of responsibility. The TSOC commander is also the permanent theater joint force special operations component commander. He commands the TSOC and is the principal special operations advisor to the GCC. The TSOC is a subordinate command of a combatant command or a functional component command of another permanent joint command. For example, the Special Operations Component, U.S. Pacific Command, is a subordinate command of U.S. Pacific Command. Special Operations Command Korea is a functional component command of U.S. Forces Korea, itself a subordinate command of U.S. Pacific Command. To provide the necessary unity of command, each GCC has established a TSOC as a subordinate command. Figure 5-3 shows the locations of the TSOCs supporting GCCs worldwide.

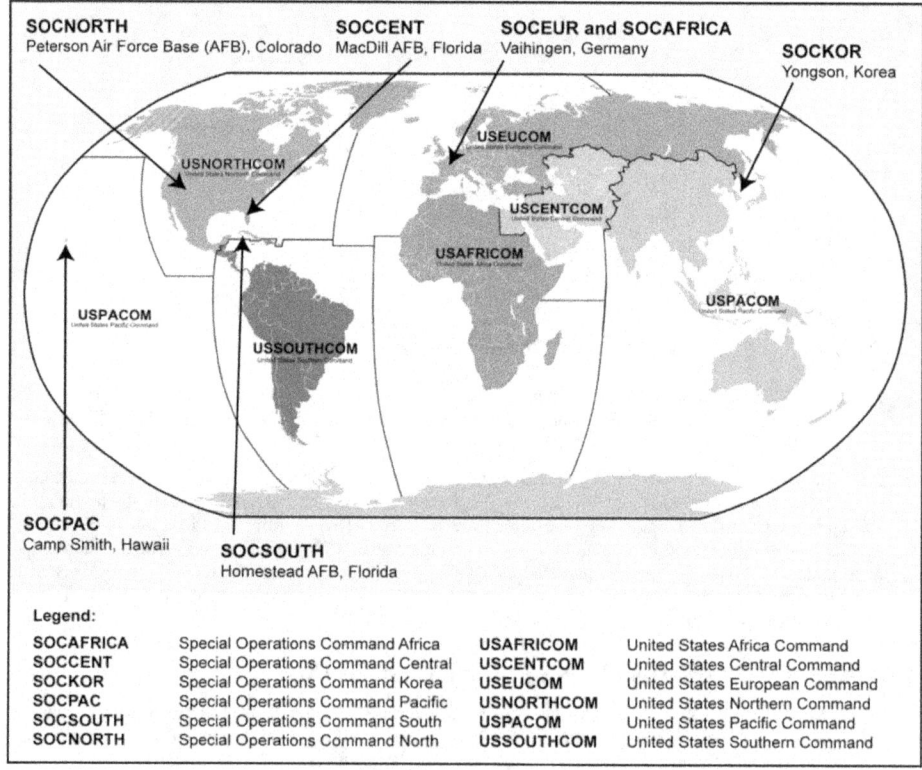

Figure 5-3. Theater and Korean Peninsula special operations commands

5-11. The TSOC is the primary theater special operations organization capable of performing broad, continuous missions well-suited to special operations capabilities. The TSOC commander has three principal roles:

- **Joint Force Commander.** As the commander of a subordinate combatant command, the TSOC commander is a JFC. As such, the commander has the authority to plan and conduct joint operations as directed by the GCC and to exercise operational control of assigned commands and forces, as well as attached forces. The commander of the TSOC may establish JTFs that report directly to him, such as a JSOTF, to plan and execute these missions.
- **Theater Special Operations Advisor.** The commander of the TSOC advises the GCC and the other component commanders on the proper employment of SOF. The commander of the TSOC may develop specific recommendations for the assignment of SOF in the theater and opportunities for SOF to support the overall geographic combatant command campaign plan. The role of the theater special operations advisor is best accomplished when the GCC establishes the commander of the TSOC as a special staff officer on the theater staff (in addition to duties as a commander—that is, "dual-hatted"). In this case, the commander of the TSOC may appoint a deputy as his representative to the theater staff for routine day-to-day staff matters.
- **Joint Force Special Operations Component Commander.** When designated by the GCC, the commander of the TSOC functions as a joint force special operations component commander. This situation normally occurs when the GCC establishes functional component commanders for operations without the establishment of a JTF. The commander of the TSOC can also be designated the joint force special operations component commander within a JTF if the scope of the operations conducted by the JTF warrants it. The joint force special operations component commander is the commander within a combatant command, subordinate combatant command, or JTF responsible to the establishing commander for making recommendations on the proper employment of SOF and assets, for planning and coordinating special operations, or for accomplishing such operational missions as may be assigned. The joint force special operations component commander is given the authority necessary to accomplish missions and tasks assigned by the establishing commander. The commander of the TSOC or the commander of the special operations JTF is normally the individual functioning as a joint force special operations component commander. When acting as a joint force special operations component commander, the individuals retain their authority and responsibilities as JFCs. A joint force special operations component commander may command a single JSOTF or multiple JSOTFs. If there is more than one JSOTF to command, the commander of the TSOC is normally established as a special operations JTF. If only one JSOTF is established (for example, within a JTF), the commander, JSOTF, may be dual-hatted as the joint force special operations component commander. When a joint force special operations component is established and combined with elements from one or more coalition nations, it becomes a combined forces special operations component and its commander becomes a combined forces special operations component commander.

5-12. Commanders of TSOCs functioning as subordinate commanders are typically established on a functional (versus geographic) basis. They normally exercise operational control of attached forces within their functional (special operations) area. They may choose to organize subordinate forces in accordance with JP 3-0 along Service and functional lines or as subordinate JTFs. The TSOC, like all joint forces, includes Service forces. Administrative and logistics support is provided through these Service forces.

5-13. The TSOCs, organized as subordinate combatant commands (all except Special Operations Command Korea), are permanently organized along Service component lines. Each TSOC has an Air Force SOF commander, ARSOF commander, and a Navy SOF commander. The Air Force provides an Air Force special operations group or wing to serve as the Air Force SOF component of the TSOC. The Navy provides a naval special warfare task group to serve as the Navy SOF component of the TSOC. The Army provides an Army SF battalion or group to serve as the ARSOF component of the TSOC. These component commanders may provide centralized operational control of assigned and attached forces for the commander of the TSOC, or they may provide forces to a functionally organized component or JSOTF subordinate to the TSOC. They may also be directed to attach forces to another organization external to the TSOC (for example, a JTF) for employment by that JFC.

Chapter 5

5-14. The TSOC may be assigned a geographic area for conduct of special operations. This geographic area is designated a joint special operations area. All external agencies must coordinate with the TSOC or special operations JTF before conducting any actions that may affect operations within the joint special operations area. In addition to U.S. SOF, other U.S. or multinational military organizations and interagency organizations may have elements operating in the joint special operations area. The commander of the TSOC makes every effort to identify these elements. He coordinates to establish command and support relationships and establishes the proper degree of coordination and cooperation through liaison elements.

DISTRIBUTIVE COMMAND AND CONTROL

5-15. The post-9/11 shift away from the conventional to irregular warfare has dramatically affected the national policy and military strategy, which, in turn, has caused both expansion and evolution of the U.S. special operations global effort. This change has been experienced on differing scales in every geographic combatant command area of responsibility and affected every TSOC. TSOCs are developing a theater-wide architecture of distributive command and control in which command nodes of varying sizes are inserted into regional campaigns to provide a special operations voice and influence to the JTFs and Chiefs of Mission. Such an arrangement improves SOF relationships with other agency partners and allies, and it helps the TSOC develop a cadre of campaign experts who rotate between nodes and TSOC headquarters. Figure 5-4 is an example of distributive command and control.

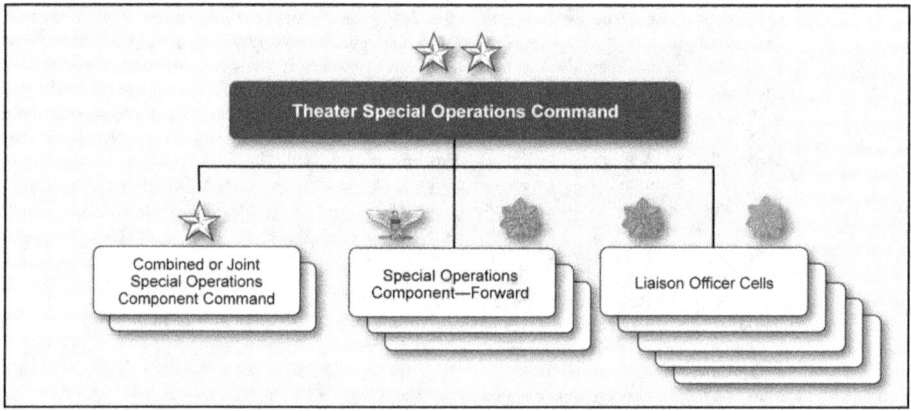

Figure 5-4. Example of distributive command and control

5-16. Distributive command and control provides continuity and leadership in key locations where SOF operate, and nodes that facilitate the surging (with force economy) of high-demand low-density enablers, if needed. Additionally, distributive command and control enables special operations to better integrate campaign efforts with JTF headquarters and to coordinate with Embassies and HN organizations. Further, it relieves tactical elements from policy-level engagement tasks so that they can concentrate on the tactical matters that are their primary purpose. Distributive command and control allows the TSOC to remain postured to leverage its platforms in places where forces are deployed and support GCC requirements where the forces cannot be deployed. Finally, distributive command and control ensures that the assigned effects are achieved in country proactively and preventatively. Such a posture is the one and only capability that can touch the whole region.

5-17. The following architecture's conceptual modules are used where SOF-centric campaigns within specific lines of operations are being executed. These modules can be grouped into the following three categories:

- **Special Operations Forces Component Command.** This module exists when a JTF is present; for example, the Combined Forces Special Operations Component Command-Afghanistan. The SOF component command is the largest of the three types of modules and is sufficiently functional in order to interface properly with a JTF on the JTF's full battle rhythm. The SOF component command will be under the command of a SOF general officer and operate in nations where DOD leads U.S. Government efforts. The primary goals of the SOF component command are to defeat the enemy, provide security, and conduct the full range of military operations in concert with conventional forces. By demonstration in both Iraq and Afghanistan, relieving the JSOTF of this interface requirement provides dividends at all levels of the chain of command.
- **Special Operations Component—Forward.** These modules have been employed in key strategic locations outside of major combat theaters/countries where there is persistent operational activity, but no JTF. The special operations command-forward is literally a persistent extension of the TSOC headquarters into the HN, providing a high confidence conduit for the commander of the TSOC to partner with the U.S. country team maintaining direct operational oversight, while providing dedicated leadership and continuity with a direct connection to the chief of mission/embassy. Normally led by an experienced SOF Army colonel or Navy captain, the special operations command-forward is empowered by the commander of the TSOC with decisionmaking authorities to proactively engage with HN and U.S. mission organizations. A special operations command-forward will operate in nations where the Department of State leads U.S. Government efforts. In these instances, both the HN and U.S. mission are sensitive to the footprint in size and visibility of a SOF presence in the HN. The primary goals of the special operations command-forward are to build relationships, provide situation awareness, coordinate operations, and apply indirect approach.
- **Liaison Officer Cells.** These are used where operational activity is recurring, but still episodic and of a smaller scale; thus, the only persistent requirement in country is for a liaison function. Normally led by an experienced SOF Army lieutenant colonel or Navy commander, these liaison officer cells will operate in the theater where the Department of State leads U.S. Government efforts. Here, both the HN and U.S. Government are very sensitive to footprint in size and visibility of a SOF presence. The primary goals are to build relationships, provide situation awareness, and facilitate planning.

5-18. The architecture's nodes are tethered by design and have operational reachback to the TSOC headquarters. Use of this reachback and a rotation of personnel between the forward and rear nodes is a SOF variation of the Afghanistan/Pakistan Hands Program. Reachback allows the TSOCs to capitalize on regional experience and grow functional depth or "benches" of experienced personnel within many disciplines. This has nurtured habitual relationships with their functional counterparts in a number of countries, in addition to having operational relationships peculiar to one specific country/location. Such reachback can be efficiently achieved if, among other things, it is enabled by an appropriate distribution of high-demand low-density specialties that cannot reasonably be populated to each node. For that reason, it is logical to centralize a number of capabilities at TSOC headquarters, within reach, and then make these capabilities available, as required, to the forward nodes on a case-by-case basis. Examples of this include intelligence analysts, engineer project managers, and SOF and contingency contracting officers. Special Operations Component, U.S. Central Command, distributive command and control nodes, as well as those similar nodes employed by other TSOCs, are essential to the current execution of the command's SOF campaign plans.

5-19. Successfully achieving the engagement missions assigned to special operations requires strong relationships with interagency, allies, and partners that are best enabled and maintained through presence in the region and sustained, focused interaction with the partner forces. Distributive command and control is the primary method for controlling and synchronizing widespread and simultaneous operations, activities, and actions. Each node provides the means to influence complex environments through the development of a long-term relationship with partner-nation leaders and the American Embassy country teams. This capability promotes regional solutions to specific problems sets. It also extends operational reach, enhancing capability to conduct campaign planning, and provides a special operations organizational platform that can rapidly transition from a peacetime to wartime footing in times of crises to execute contingency plans.

TASK FORCES

5-20. Each GCC has tasked specified subordinate commanders with the "be prepared" mission of forming a JTF headquarters to support area of responsibility requirements (Figure 5-5). The JTF headquarters may be established for a short-duration mission, such as a contingency operation (for example, noncombatant evacuation operation), or for continuous joint operations, such as JTF-Bravo in Honduras.

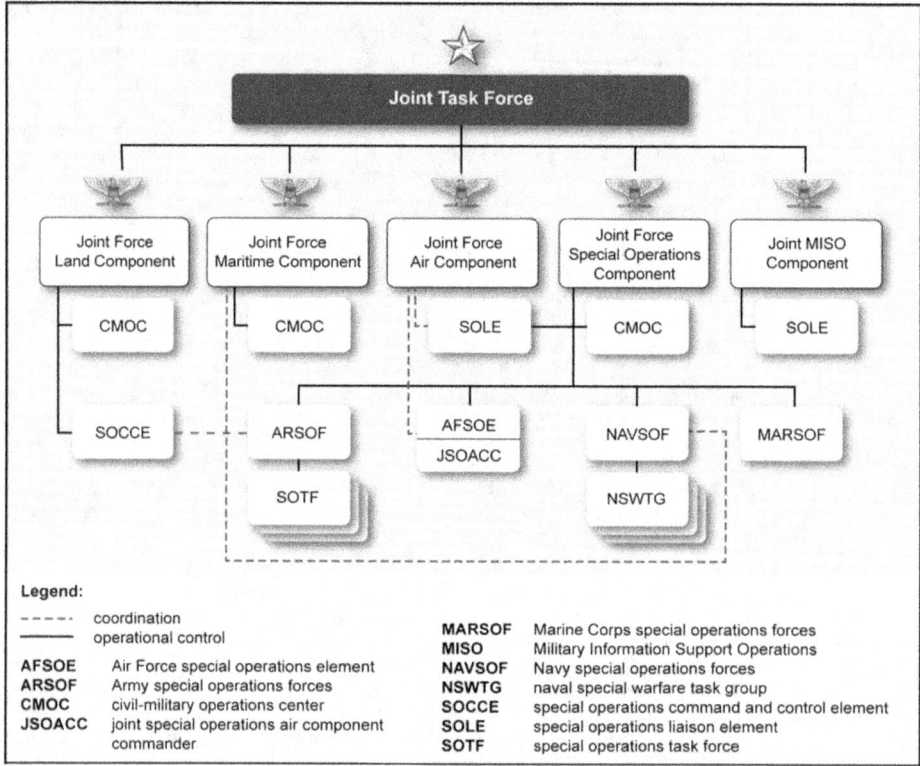

Figure 5-5. Notional joint task force command and control

5-21. Headquarters, USSOCOM, established a deployable JTF headquarters core staff (JTF 487). JTF 487 achieved an initial operating capability as of 31 May 2010. JTF 487's organizational construct provides the GCC a capability optimized for irregular warfare. It is structured and equipped to deploy within 10 days of notification in support of a GCC's emerging irregular warfare JTF requirements. It is ideally suited to substitute for, sustain, or replace a deployed TSOC or other organization. JTF 487 combines three USSOCOM elements:

- **Joint Task Force Sword.** This task force provides the first echelon of the JTF headquarters staff and within 10 days establishes an initial command and control operating capability (18 to 31 personnel).

Employment

- **Headquarters, United States Special Operations Command.** This headquarters provides key staff personnel to establish the second echelon within 30 days and brings joint assets and seasoned irregular warfare experienced staff.
- **U.S. Army Special Forces Command.** This command provides the commanding general and the command element for JTF 487. Commander, USASFC, task organizes based on mission analysis to ensure specific skill sets are tailored for the GCC, the country, and the mission, and must be prepared to deploy within 30 days of notification.

5-22. JTF Sword is a headquarters with the mission tasking of forming a JTF. JTF Sword is a joint table of allowances and joint table of distribution organization for the purpose of providing the Commander, USSOCOM, the ability to deploy a special operations JTF core element to facilitate the rapid establishment of a command and control element to support GCC requirements. JTF Sword is an organized, trained, and equipped deployable SOCCE that will primarily execute missions at the operational level for short-duration deployments (less than 180 days). JTF Sword is also tasked by Commander, USSOCOM, to test and conduct combat and operational evaluations of SOF control systems. JTF Sword is organic to USSOCOM and assigned under the USSOCOM Directorate of Operations with duty at Fort Bragg, North Carolina. JTF Sword is the primary core element for USSOCOM's JTF 487. Commander, USSOCOM, can also task JTF Sword to deploy independently of JTF 487.

5-23. The TSOC is another one of the headquarters with this mission tasking. It provides the GCC a responsive special operations JTF headquarters capability for contingencies. It is proficient in joint, multinational, and interagency operations. In those situations where another subordinate is tasked to form the JTF, the TSOC recommends an appropriate special operations force structure for attachment to, and employment by, that JFC.

5-24. As described in JP 3-0, a JFC normally organizes forces with a combination of Service and functional components and subordinate JTFs. Although all joint forces include Service components, a key decision by the JFC is the type of functional components or JTFs required to accomplish the mission. In the SOF environment, these two organizations are the joint force special operations component (a functional component) or a special operations JTF (a subordinate JTF). A joint force special operations functional component lacks the organizational flexibility and authority of a JTF. Therefore, JFCs typically establish a special operations JTF.

Special Operations Joint Task Force

5-25. The special operations JTF is a task force through which USSOCOM will present all theater SOF under one special operations commander. For crisis response, contingency, and major operations and campaigns, SOF may deploy a special operations JTF where all SOF report to one special operations commander and the packaged force includes all enabling capabilities (organic to special operations formations and those Service-provided conventional forces capabilities) required to optimize the effectiveness of the special operations JTF. A special operations JTF is an operational-level organization that may have one or more subordinate JSOTFs (Figure 5-6, page 5-10).

Joint Special Operations Task Force

5-26. When deployed, SF operations are organized by a system of operational bases into task forces—the JSOTF, the SOTF, and the AOB. Each level of SF command task organizes its task force to establish tactical facilities at these operational bases with specific functions.

Chapter 5

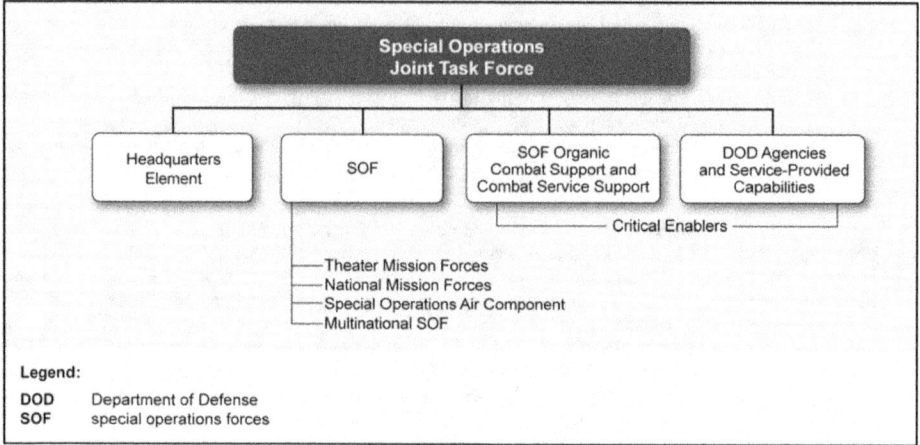

Figure 5-6. Notional theater special operations joint task force

5-27. A JSOTF is a JTF composed of special operations units from more than one Service established to conduct a specific special operation or to prosecute special operations in support of a campaign or other operations. A JSOTF may have conventional non-special operations units assigned or attached to support the conduct of specific missions in the following situations:

- A JSOTF, like any JTF, is normally established by a JFC—for example, a combatant commander, a subordinate combatant commander (such as a commander of a TSOC), or a JTF commander. For instance, a GCC could establish a JTF to conduct operations in a specific region of the area of responsibility. Then, either the GCC or the JTF commander could establish a JSOTF, subordinate to that JTF, to plan and execute special operations. Likewise, a commander of a TSOC could establish a JSOTF to focus on a specific mission or region assigned by the GCC. A JSOTF may also be established as a joint organization and deployed as an entity from outside the area of responsibility.
- A JSOTF is established to conduct operations in a specific area of operations or to accomplish a specific mission. If geographically oriented, multiple JSOTFs are normally assigned different areas of operations.
- Within a JTF, if only one JSOTF is established, the commander, JSOTF, is dual-hatted as the joint force special operations component commander. When a JSOTF is established to support a GCC directly, the commander of the TSOC normally acts as the JSOTF commander. Regardless, a JSOTF commander is a JFC and exercises the authority and responsibility assigned by the establishing authority. A JSOTF staff is normally drawn from the TSOC staff or an existing SOF component with augmentation from other SOF or conventional units and personnel, as appropriate.
- When a JSOTF is established and combined with elements from one or more coalition nations, it becomes a CJSOTF and its commander becomes a CJSOTF commander.

5-28. A JSOTF headquarters can operate under a number of command relationships. It may be established to conduct a specific special operation or to prosecute special operations in support of an operational campaign. As stated in JP 3-0, a JFC (either a GCC or subordinate special operations JTF commander) may organize the geographic area and forces in any manner to best accomplish assigned missions. The JFC may direct the JSOTF commander to support another component commander, or subordinate the JSOTF under another JFC (for example, a special operations JTF commander). The JFC may also attach other Service forces under the control of the JSOTF for the conduct of operations. The headquarters Service composition depends on the mission, operational environment, available capabilities and support, and composition of forces. The JSOTF headquarters may be sourced from USSOCOM and other Service component assets, the TSOC headquarters, or special operations command components.

> **Combined Joint Special Operations Task Force-Afghanistan**
>
> The CJSOTF-Afghanistan was established in Bagram, Afghanistan, in May 2002. CJSOTF-Afghanistan was predominantly manned by personnel from the headquarters elements of the SFG that was currently conducting combat operations in Afghanistan. Personnel from other joint and combined branches of the armed forces completed the manning for this unique organization. Once fully operational, CJSOTF-Afghanistan assumed responsibility for the command and control of most SOF in Afghanistan.
>
> When established, CJSOTF-Afghanistan consisted of two U.S. Army SF battalion SOTFs along with the SOF task forces from several partnered nations, including the United Arab Emirates, France, Australia, Canada, and the United Kingdom. As the mission evolved over time to support strategic objectives, CJSOTF-Afghanistan had grown in size and (as of 2011) consisted of two U.S. Army SF SOTFs, one U.S. Marine Corps SOTF, and one U.S. Navy Special Warfare SOTF along with United Arab Emirates, Canadian, and French SOF task forces.

5-29. As noted, ARSOF may be tasked to form the core of an ARSOF-heavy JSOTF. The SFG is the most likely ARSOF headquarters to form this core. The group headquarters is the preferred ARSOF headquarters because of its organic mission command capabilities. The group has self-contained communications and support elements. It also has area expertise, extensive experience with and knowledge of instability, and prior operations with the foreign nation government and its military forces. However, significant augmentation from the other Services is required to transform any ARSOF headquarters into a JSOTF. Figure 5-7, page 5-12, depicts a notional JSOTF.

COMBINED FORCES SPECIAL OPERATIONS COMPONENT COMMAND

5-30. The combined forces special operations component command is an emergent special operations headquarters element that has been used to provide special operations representation to interface with appropriate campaign command and control elements. This headquarters represents theater special operations and their with-and-through perspective in HN senior leader-level engagements, and in the U.S./coalition boards, centers, and cells that help shape the operational campaigns, which are in many ways very SOF-centric (Figure 5-8, page 5-13).

Special Operations Task Force

5-31. The JSOTF may consist of SOF from each of the Services. Each Service force senior commander retains Service responsibilities (administrative control) of those Service forces. For example, in most instances, the SFG commander exercises administrative control of Army personnel attached to the JSOTF. Likewise, the Air Force SOF commander exercises administrative control of Air Force personnel.

Chapter 5

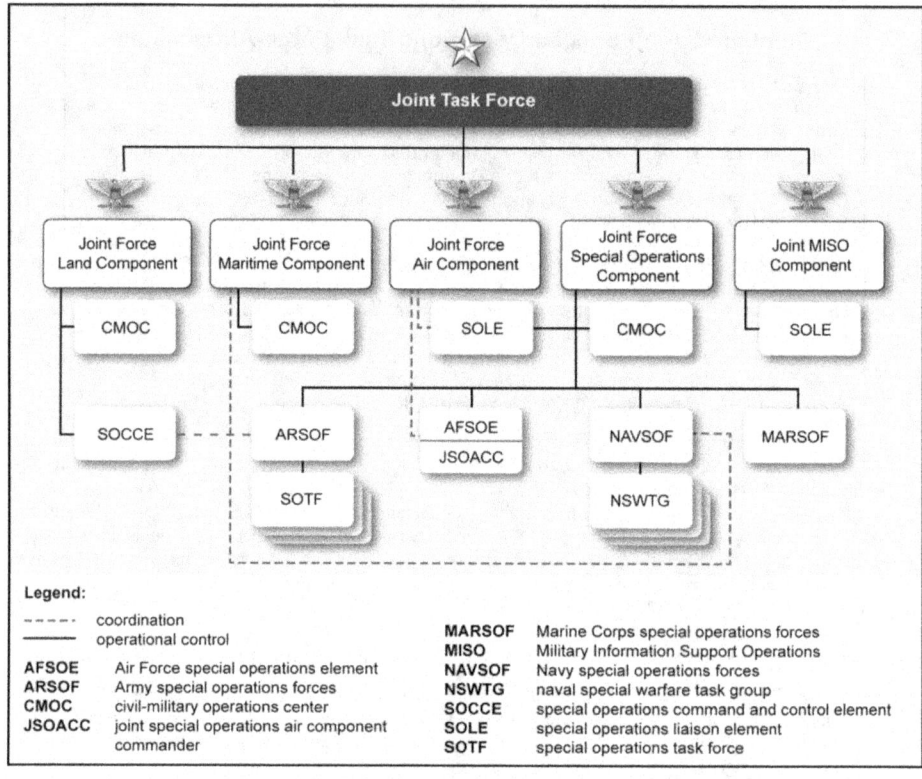

Figure 5-7. Notional joint special operations task force

Theater Special Operations Forces Expertise

In the Special Operations Component, U.S. Central Command, area of responsibility (Iraq and Afghanistan) both had ample theater SOF expertise at the tactical level in the form of the CJSOTF. These formations are the best at what they do; however, what was missing was theater SOF expertise at the operational level, namely at the International Security Assistance Force and U.S. Forces-Iraq.

To fill the void, two one-star-level commands were fielded: Joint Force Special Operations Component Command-Iraq and Combined Forces Special Operations Component Command-Afghanistan.

Employment

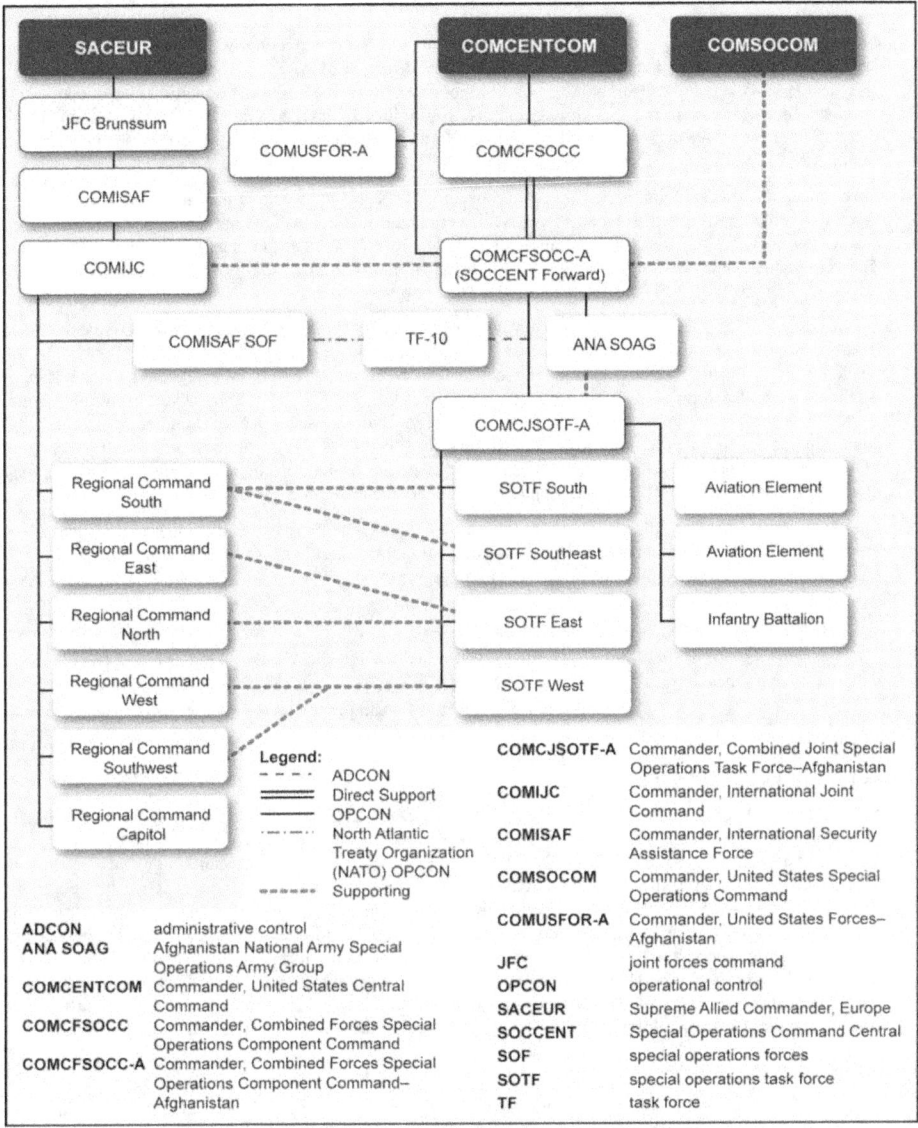

Figure 5-8. Example of Combined Forces Special Operations Component Command—Afghanistan organization

5-32. The JSOTF commander may also designate and organize operational and functional organizations for employment of forces. The commander may designate and organize a joint special operations air component commander to control all special operations air assets functionally. Likewise, a single SOTF may be designated and organized to provide operational direction of ARSOF. The JSOTF commander may also decide to designate and organize several subordinate SOTFs to conduct specific special operations missions. In this situation, the JSOTF commander would directly exercise operational control of each task force and the senior ARSOF commander would continue to exercise administrative control responsibilities for all ARSOF within those task forces.

5-33. When the JSOTF commander has numerous and diverse missions and large numbers of Army forces, he may designate multiple SOTFs and exercise direct operational control of each. Each SOTF is organized around the nucleus of an ARSOF unit and can include a mix of ARSOF units and their support elements. The commander, JSOTF, assigns each SOTF an area within the joint special operations area or functional mission. Figure 5-9, page 5-15, shows a notional SOTF organization.

5-34. Since the SFG and SF battalion are multipurpose and extremely flexible organizations designed to have self-contained mission command and support elements for long-duration missions, the SOTF headquarters is normally based around the core of an SFG or SF battalion headquarters. The Ranger Regiment may also form a SOTF headquarters for large-scale Ranger operations. The SOTF commander augments a staff with appropriate special staff officers and liaison officers taken from attached and supporting assets to integrate and orchestrate all activities of the SOTF. In the case of an extremely large SOTF that exceeds the mission command capabilities of the organic headquarters, the commander may have to request external staff augmentation to ensure adequate mission command.

5-35. In some situations, the SOTF or a subordinate ARSOF unit may receive operational control or attachment of a conventional maneuver unit. This situation most likely occurs—

- In a UW environment when an ARSOF-supported indigenous combat force needs added combat power for a specific combined arms operation.
- When the SOTF needs a conventional reaction or reinforcement force for its special operations.
- In linkup or post-linkup combat operations during the combat employment phase of an insurgency.
- During contingency operations when the SOTF headquarters is the senior Army headquarters in the area of operations.

TASK FORCE ORGANIZATION

5-36. The SOTF commander will normally task organize a unit into a series of centers to coordinate operations when deployed. A center is a control facility with a supporting staff established for a specific purpose. These centers have a narrow focus; however, the activities of these centers are interdependent. The centers are normally the operations center, the signal center, and the support center. The operations center revolves around operational concerns, from mission planning to intelligence and execution. The signal center installs, operates, and maintains the unit's internal and external systems. The support center provides logistics support to the unit. These centers are also normally formed at the battalion or SOTF level to mirror the group. Although the battalion SOTF or group JSOTF may form these centers, the staff sections for these organizations are still very adept at communicating, functioning, and interacting with the staff sections of both conventional or coalition forces that are organized in a standard staff configuration.

Operations Center

5-37. Whereas functional cells group personnel and equipment by warfighting functions, integrating cells group personnel and equipment to integrate the warfighting functions by planning horizons. A planning horizon is a point in time commanders use to focus the organization's planning efforts to shape future events. Planning horizons are situation-dependent; they can range from hours and days to weeks and months. As a rule, the higher the echelon, the more distant the planning horizon with which it is concerned.

Employment

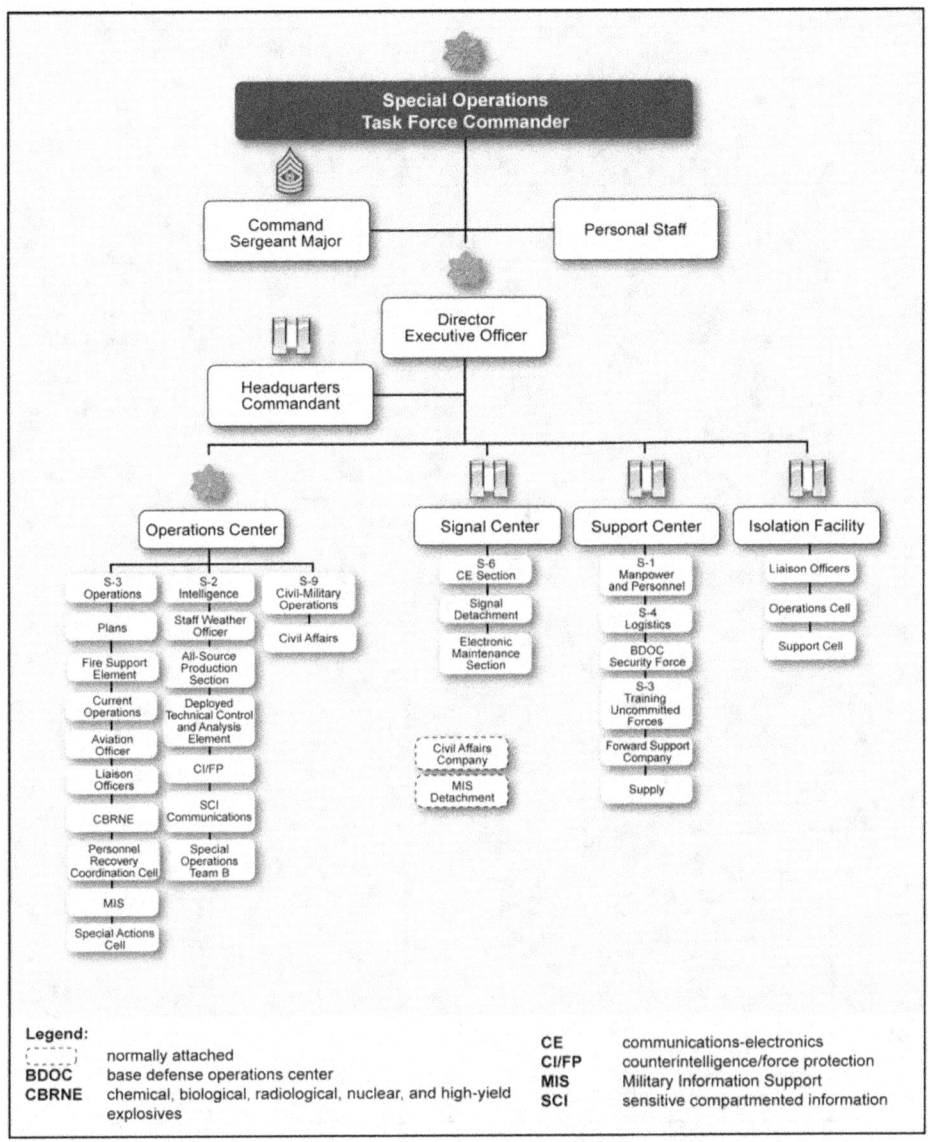

Figure 5-9. Notional special operations task force organization

5-38. The operations center is cross-functional by design and, by including representatives from almost all staff sections within the SOTF, supervises all operational elements of mission planning and execution. Combined with a robust communications package, the operations center is charged to plan, prepare, coordinate, execute, control, and assess all current, mid-range, and future operations for a designated area of responsibility. It performs the functions of a conventional unit's tactical operations center. Figure 5-10, page 5-16, shows the operations center organization.

Chapter 5

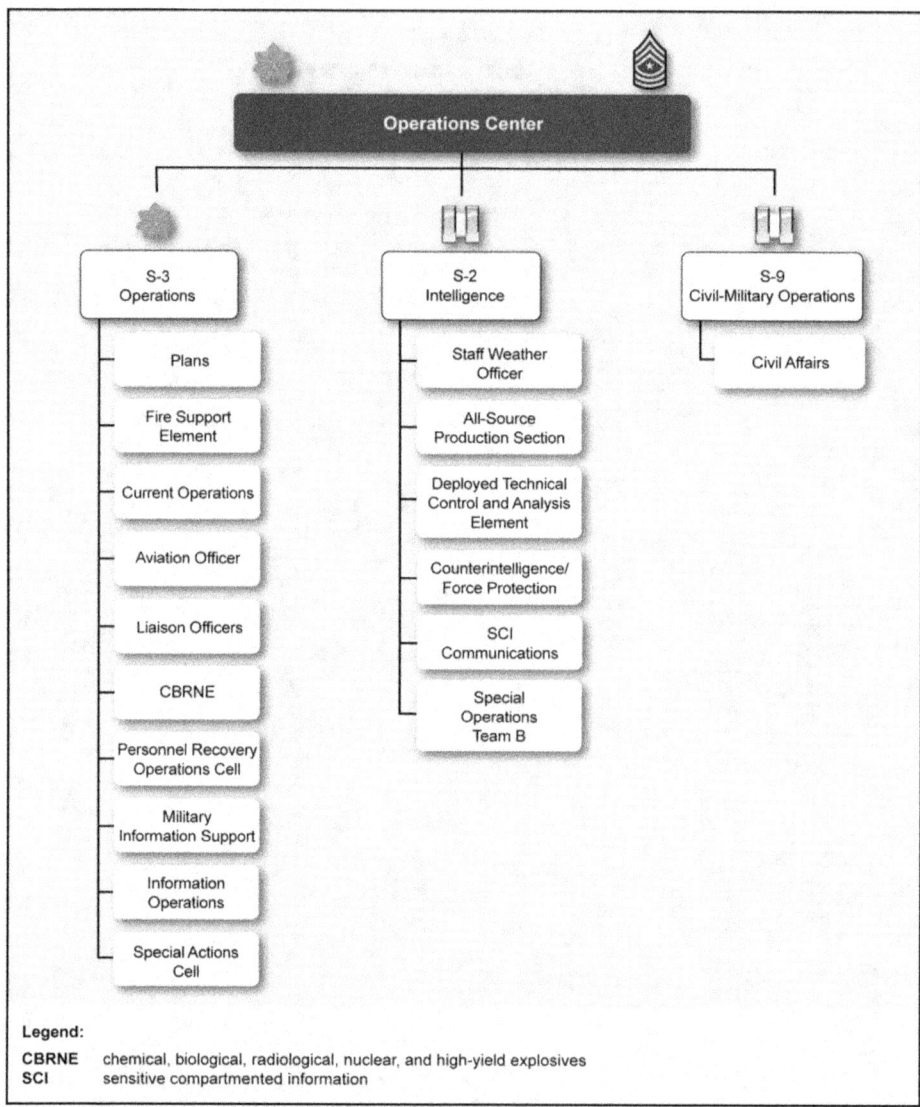

Figure 5-10. Operations center organization

5-39. The operations center director, normally the task force operations staff officer, ensures the complete orchestration of the intelligence staff office, operations staff office, and CAO staff office activities within the operations center. The operations center director or SOTF commander may further choose to task organize the operations center into smaller sections or cells dependent on the current situation and mission.

Examples include the joint operations center, the fusion cell, the nonlethal effects cell, the special actions cell, and the future operations cell. These centers or cells are formed and manned to perform specific functions within the operations center, such as current operations, plans, and future operations and are described below. Figure 5-11 shows a notional operations center.

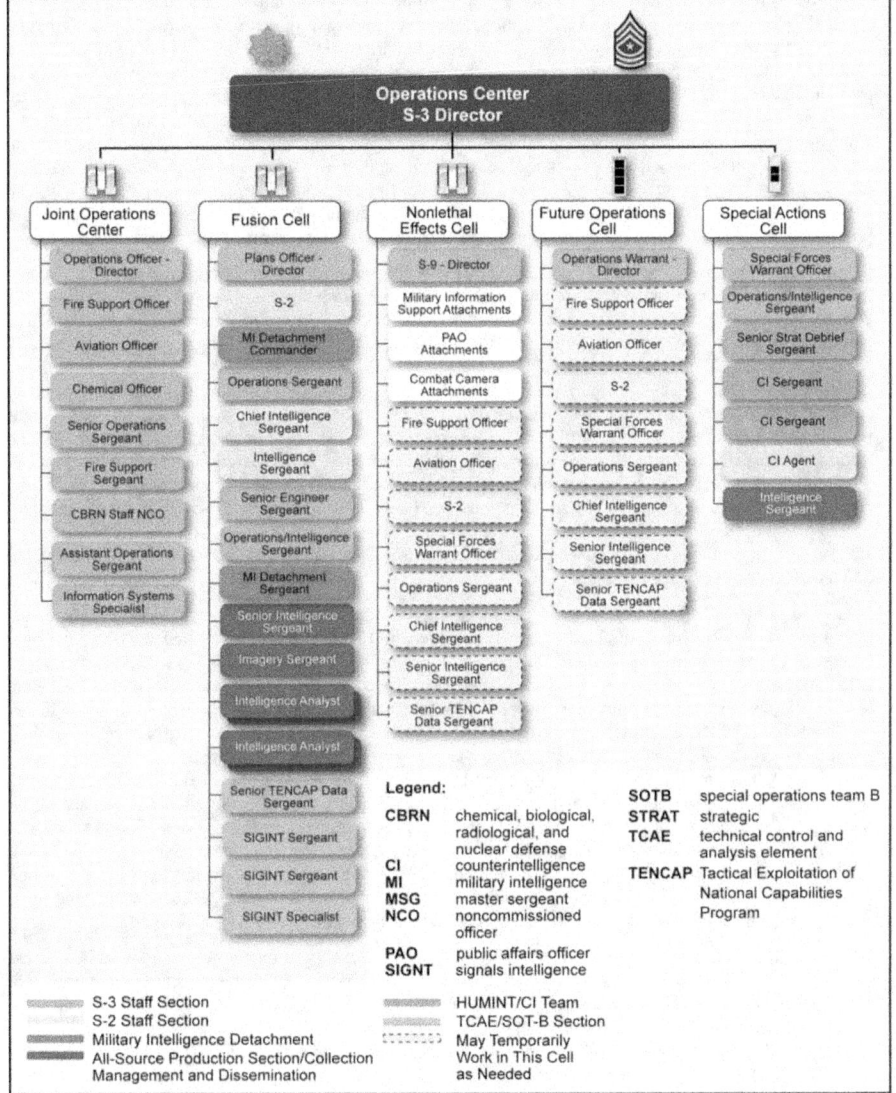

Figure 5-11. Notional operations center

5-40. The joint operations center will normally be directed by the task force current operations cell, and is the focal point for all operational matters. The joint operations center will be charged to monitor, direct, coordinate, and control all current operations and taskings. The planning horizon for the joint operations center will be situational dependent and determined by the operations center director or SOTF commander. It may range from hours to days. Normal manning usually consists of the operations officer or assistant operations officer as the battle captain, the aviation officer, the fire support officer, the chemical officer, the senior operations sergeant, the fire support sergeant, the assistant operations sergeant, the information systems specialist, the special operations weather team, and the various liaison officers for higher, adjacent, subordinate, supporting, and supported units operating in the designated area of responsibility. In order to provide the task force commander with all critical and relevant information during special events or critical situations, the joint operations center may also be further manned or supported with additional personnel from the support center, fusion cell, nonlethal effects cell, or special actions cell, such as the medical section, personnel staff office, the judge advocate general, the MIS officer, the CAO officer, and intelligence or imagery analysts as necessary. This will allow the joint operations center, fusion cell, and nonlethal effects cell to perform crisis or time-sensitive planning and determine when to implement potential branches and sequels as current plans and operations develop to ensure success on the battlefield.

5-41. The fusion cell will normally be directed by the task force future operations cell, and will be charged to monitor all activities on the battlefield and conduct the detailed resourcing, training, manning, and equipping needed for future plans and future operations that will drive the fight on the battlefield. The planning horizon for the fusion cell is also situational dependent and may range from days to weeks. Normal manning will usually consist of the future operations cell, intelligence staff officer, military intelligence detachment commander, operations sergeant, operations and intelligence sergeant, engineer sergeant, intelligence analysts, imagery analysts, counterintelligence, human intelligence, and signals intelligence personnel. By performing detailed planning, close coordination, synchronization, and information sharing across staff sections according to lines of effect and lines of operations, the fusion cell is able to develop effective plans, branches, and sequels that achieve the task force commander's desired results. During special events or critical situations, the fusion cell may support the joint operations center by providing the task force commander with current and historical intelligence information for a specific area or specific individual, as well as the ability to control and analyze full-motion video collected from unmanned aircraft systems.

5-42. The nonlethal effects cell may be established within the fusion cell, or as its own cell under the operations center. Normally directed by the task force CAO staff officer, the nonlethal effects cell will consist of the task force CAO staff office, public affairs officer, combat camera personnel, and attached MIS. Additional personnel from the joint operations center, fusion cell, or special actions cell may temporarily work within the nonlethal effects cell as the situation or mission may dictate. These personnel will contribute their subject-matter expertise to the nonlethal effects cell mission and then return to their normal duties. The nonlethal effects cell will be charged with contributing information superiority to plans by promoting current and future operations and by degrading or negating the threat's operations, which results in a distinct information advantage. This is achieved by integrating information-related capabilities into the early stages of developing operation plans, concepts of operations, and operations orders, and applying their capabilities across the full range of operations. Examples include overseeing various civil-military operations projects and military deception operations, conveying the approved U.S. or coalition task force message to the local populace, coordinating for both U.S. and foreign media events and press releases, and managing any positive or negative consequences resulting from current operations.

5-43. The future operations cell is effectively the program manager used for future planning needed for the expansion of current operations and programs. This cell provides the commander a wider range of possibilities or courses of action to make more of an influence on the ever-changing world and battlefield to come. The planning horizon for future operations may range from weeks to months. With only a small, dedicated cell of core planners under the direction of the operations warrant officer, all staff sections, centers, and cells assist as required, and then return to their normal duties. Whereas a JSOTF has a small group of core planners, a SOTF is not fully resourced for a plans cell and personnel balance future operations with other staff functions.

5-44. The special actions cell may be established within the fusion cell or as its own cell under the operations center. *Special actions* are those functions that, due to particular sensitivities, compartmentation, or caveats, cannot be conducted in normal staff channels and therefore require extraordinary processes and procedures and may involve the use of sensitive capabilities (JP 3-05.1). These actions are frequently interagency in nature, involve sensitive capabilities, and almost always are not releasable to coalition partners or allies. The special actions cell, under the direction of the SF warrant officer or a plans officer, may be manned with the operations and intelligence sergeant, the senior strategic debrief sergeant, counterintelligence personnel, and the intelligence sergeant. The special actions cell does not control intelligence functions that are normally purview to the intelligence staff officer, but coordinates special actions closely with the intelligence staff officer functions. Generally, the special actions cell will be responsible for the following functions:

- Compartmented planning support.
- Interagency/Special Mission Unit operational integration.
- Advanced special operations.
- Special operational support.
- Special tactical operations.
- Nonconventional assisted recovery.
- Compartmented information management.

Further description and information of the special actions cell and its responsibilities is found in JP 3-05.1.

Signal Center

5-45. Under the direction of the task force signal staff officer, the signal center installs, operates, and maintains secure, reliable, and long-range communications between the task force and higher, adjacent, subordinate, supporting, and supported headquarters, and deployed ODBs and ODAs. The signal center also installs, operates, and maintains internal and external communications. The signal center normally consists of the unit signal staff section, organic signal detachment, and attached or supporting signal elements (Figure 5-12). At the JSOTF level, the signal center director normally task organizes these assets into a signal support coordination element, electronic maintenance section, system control section, communications security section, communications center, multimedia section, and message center. In most instances, no multimedia function exists at the SOTF level and the system control section may also perform signal support coordination element functions. Additionally, the SOTF does not have an organic electronic maintenance section and normally evacuates electronic equipment to the JSOTF for maintenance.

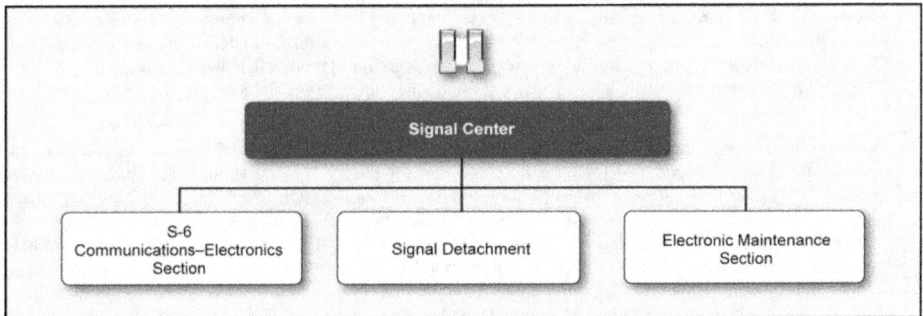

Figure 5-12. Signal center organization

Chapter 5

Support Center

5-46. When the group establishes a JSOTF, the group support operations officer serves as the support center director and the headquarters support company commander assumes these duties when the battalion forms a SOTF. These duties require direct interface with theater Army logistics support elements. In coordination with the operations staff officer and headquarters commandant, the group support company commander prepares the base defense plan and supervises the activities of the JSOTF base defense operations cell.

5-47. The support center is the functional activity that provides logistical support to the task force and its subordinate elements. The support center performs the functions of a conventional unit's trains. The support center normally consists of the task force personnel staff office, logistics staff office, forward support company, motor pool, medical section, property book office, unit ministry team, organic support company, and appropriate direct support-level logistics and force health protection attachments (Figure 5-13). The support center director is normally the support company commander, who is responsible to the task force commander for the execution of all base logistics support and force health protection operations, base support plan, and base security.

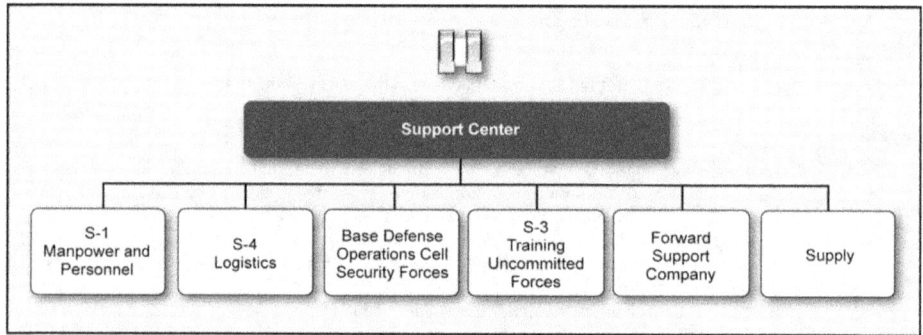

Figure 5-13. Support center organization

5-48. The HHC commander or the group support company commander can be the headquarters commandant at the JSOTF level. The support company commander or the executive officer assumes the duties of the headquarters commandant at the SOTF level. The headquarters commandant is responsible for support center details, including billeting, food service, space allocation, and internal guard duties, and also directs base defense operations and base emergency evacuation planning. The headquarters commandant, along with the base chemical defense officer, establishes mission-oriented protective posture for base CBRNE defense.

5-49. At the JSOTF support center, the personnel staff office, logistics staff office, and surgeon plan and coordinate logistics support and force health protection for the JSOTF and subordinate bases. Whenever possible, they coordinate throughput distribution to the SOTF or the AOB level. When two bases collocate, the JSOTF commander may decide to collocate or consolidate the support centers for mutual support. At the SOTF support center, the personnel staff office, logistics staff office, and surgeon plan and coordinate the logistics support and force health protection for the SOTF, its subordinate AOBs, and its deployed SF units and SOT-As. At the AOB support center, logistics support is normally limited to unit-level functions that support the current operations. The support center plans and controls the administrative activities of the task force, including the coordination of facilities engineering and other base operations support. The support center maintains the unit basic load and supplies. The support center also manages stocks, coordinates movements, provides or arranges for maintenance, requisitions and coordinates logistics support requirements, controls personnel management, and supports the training and preparation of uncommitted ODAs and SOT-As.

Employment

Base Defense Operations Cell

5-50. The base defense operations cell, under the direction of the headquarters commandant, ensures specific responses and procedures are established, reviewed, and practiced for base defense. The preferred base defense option is for U.S. elements to secure the JSOTF or SOTF bases and activities. When available, military police or infantry security elements may be attached for specific base defense duties. In addition, the supported HN force may provide physical security to the base. The SF base commander's last resort option is to divert support or operational personnel for base security. SF bases employ standard CBRNE defense measures to protect themselves in the CBRNE environment.

Advanced Operations Base

5-51. Within SF, operations are normally decentralized. For this reason, an ODB establishes a small temporary base near or within a joint special operations area to mission command and/or support training or tactical operations. Facilities are normally austere. The base may be ashore or afloat. If ashore, it may include an airfield or unimproved airstrip, a pier, or an anchorage. An AOB is normally controlled and/or supported by a main operations base or a forward operations base (JP 3-05.1). Figure 5-14 shows an AOB organization.

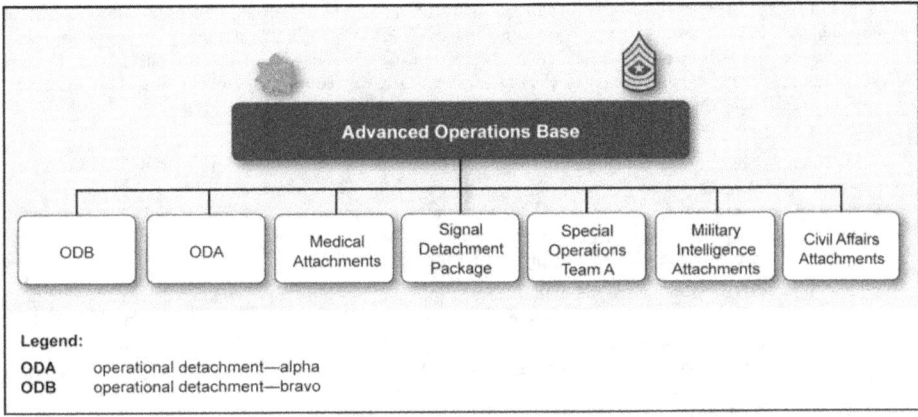

Figure 5-14. Advanced operations base organization

5-52. The ODB transformed into an AOB is a multipurpose mission command element that can exercise control over one to six ODAs. Although it will normally require augmentation from either a SOTF or JSOTF assets to perform these missions, the ODB is normally employed in one of following three ways:
- The ODB can serve as an AOB. In such instances, it is usually small, light, and tailored to perform specific missions, such as mission command, forward launch and recovery, logistics, and communications.
- The ODB can establish an isolation facility within the framework of the SOTF to isolate and prepare up to six ODAs for infiltration, mission execution, exfiltration, and reconstitution.
- The ODB can establish a SOCCE or SF liaison element at a functional component or Service force headquarters to facilitate JFC-designated command relationships between the JSOTF and that headquarters.

5-53. The ODB plans and conducts SF operations separately or as part of a larger force. It also—
- Trains and prepares ODAs for deployment.
- Acts as a mission command element of all subordinate elements during the conduct of special operations missions.

- Infiltrates and exfiltrates specified operational areas by air, land, or sea.
- Conducts operations in remote and denied areas for extended periods with minimal external direction and support.
- Develops, organizes, equips, trains, and advises or directs indigenous forces up to regiment size in special operations activities.
- Trains, advises, and assists other U.S. and multinational forces and agencies.
- Organizes, advises, and assists a UW area command in a specified joint special operations area.
- Serves as a pilot team to assess the resistance potential in a specified joint special operations area.
- Performs other special operations activities as directed by higher authority.

SF units will rarely conduct company-size operations. When multiple ODAs are tasked to conduct unilateral, joint, or multinational operations together, an ODB will usually be tasked to execute mission command and support for that specific mission.

SPECIAL OPERATIONS COMMAND AND CONTROL ELEMENT

5-54. Within a joint force, ARSOF assets are ordinarily attached to and under operational control of a designated JSOTF commander. These ARSOF assets may often operate in proximity to other components of the JTF or support those components as part of the JSOTF's mission taskings and supporting commander's responsibilities. When possible, liaison is reciprocal between higher, lower, supporting, supported, and adjacent organizations (that is, each one sends a liaison element to the other). In such instances, the JSOTF commander may elect to employ SOCCEs to coordinate unilateral special operations with conventional ground force headquarters or, if a supporting commander, facilitate his responsibilities to the supported commander.

5-55. As a supporting commander, the JSOTF commander ascertains and fulfills the needs of the supported commander within the parameters imposed by the JTF commander. The JSOTF commander determines the type of force, employment, and procedures to accomplish the support, and normally employs a SOCCE to facilitate the supporting commander's responsibilities to a ground force commander. The SOCCE remains under the operational control of the JSOTF commander.

5-56. The SOCCE assists the JSOTF commander in fulfilling the supporting commander's responsibilities in several ways. It provides a positive means for the JSOTF commander to ascertain the supported commander's needs. The SOCCE may provide a responsive reporting capability in those situations where the JSOTF commander has been requested to provide information requirements of the supported commander (for example, special reconnaissance reporting). The SOCCE can exercise command and control of designated ARSOF units when the JSOTF commander determines the need for such a command relationship to facilitate the supporting commander's responsibilities. The SOCCE can also provide a monitoring capability if the JSOTF commander decides to transfer ARSOF under a command relationship of the supported commander—for example, the attachment of SF detachments under the control (operational control or tactical control) of the Army forces to improve the Army forces commander's ability to employ subordinate multinational forces. The JSOTF commander could transfer these forces and pass control to the Army forces with appropriate mission restrictions in accordance with a determination on the employment of those forces, such as "no reorganization of forces authorized" or "for use only in an advisory assistance role with the designated multinational force."

5-57. A SOCCE is augmented with a special communications package and personnel, as required. It may include SF, Ranger, MIS, CA, special operations aviation, and other SOF representatives. The SOCCE is normally collocated at corps level and above, with smaller liaison teams operating at division level and below. The supported unit provides the SOCCE the required administrative and logistics support. The SOCCE is the focal point for synchronization with the conventional forces. At corps level, the SOCCE coordinates with the corps operations center, fire support element, deep operations coordination cell, and battlefield coordination detachment to deconflict targets and operations. It provides ARSOF locations through personal coordination and provides overlays and other data to the fires cell and the battlefield coordination detachment. Figure 5-15, page 5-23, shows a SOCCE organization.

Employment

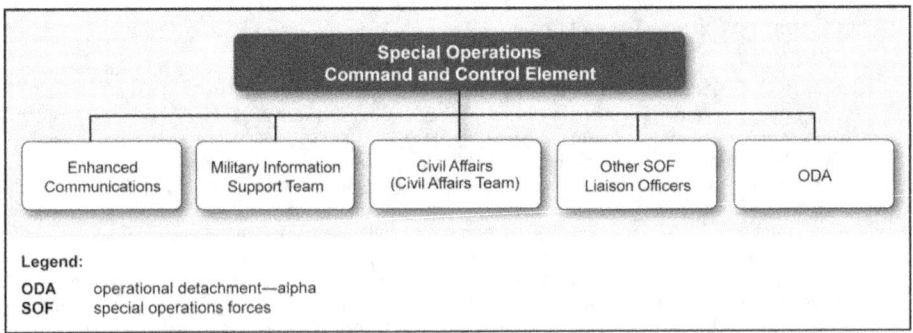

Figure 5-15. Example of special operations command and control element organization

SPECIAL FORCES LIAISON ELEMENT

5-58. The SF liaison element is an SF element that conducts liaison between U.S. conventional forces and HN or multinational forces. It is formed only as needed. SF liaison elements conduct these functions when conventional forces or host or multinational forces have not practiced interoperability before the operation, when the forces do not share common operational procedures or communications equipment, or when a significant language or cultural barrier exists.

LIAISON OFFICER MISSIONS, FUNCTIONS, AND ROLES

5-59. SF liaison officers report to the special operations commander or special operations component commander and are dispatched to applicable conventional JTF components. SF liaison officers execute liaison and coordination activities in several capacities. Whether individually, in teams, or as organic members of assigned staffs, liaison officers and coordination representatives perform several critical functions that are consistent across the range of military operations. SF liaison officers ensure the timely exchange of necessary operational and support information to aid mission execution and preclude fratricide, duplication of effort, disruption of ongoing operations, or loss of intelligence sources. They may assist in the coordination of fire support, overflight, aerial refueling, targeting, deception, MISO, and other operational issues based on ongoing and projected special operations missions. Liaison elements from SF must be trained, empowered, and resourced. Depending on the mission, it may or may not require an entire ODA. For coordination at brigade or below, ODAs may have to coordinate directly with brigade, battalion task forces, and companies, dependent upon the area of operations and the amount of coordination necessary. ODAs may need to coordinate crossing area of operations boundaries, conducting operations in specific areas of operations, or requesting support from conventional units in the area of operations in which the ODA operates.

TRAINING PROGRAMS

5-60. Training programs include theater security cooperation plan events, joint combined exchange training events, and counternarcoterrorism training.

THEATER SECURITY COOPERATION PLAN EVENTS

5-61. Many different units conduct theater security cooperation plan events covering a wide range of activities. SF units are used for specific reasons with associated laws, authorities, and fund sources. These events are tools for the GCC's peacetime campaign strategy to gain or maintain U.S. access to strategically important foreign countries in support of the American Embassy and U.S. security interests. Security cooperation activities advance those interests by creating new partnerships and building the capacity of existing partnerships. This section addresses two types of theater security cooperation plan events conducted by SF.

Joint Combined Exchange Training

5-62. The joint combined exchange training program provides combatant commanders with trained and ready SOF to execute their theater campaign plans or respond to regional crises. The joint combined exchange training program is unique in that its requirements are based on SOF mission-essential task lists, shaped by the theater strategy, orchestrated by TSOCs and the U.S. Ambassador (or his or her representative), approved by the Secretary of Defense in coordination with the Secretary of State, and executed by the individual unit. SF has unique training requirements for core missions, such as FID, counterinsurgency, and UW, in which SF detachments deploy to foreign nations in order to train on SOF mission-essential task lists and develop skills in the cross-cultural environments. Joint combined exchange training programs are primarily conducted by ODAs; however, other components of USSOCOM also participate in the program. Joint combined exchange training programs are the primary mechanism for the group commanders to train their ODAs on these unique core mission-essential task lists in their regionally aligned operational environments. The joint combined exchange training program provides regional orientation, language immersion, and combined training with potential coalition partners. Joint combined exchange training programs are also the most scrutinized form of joint and combined training and require more regulatory guidance than Joint Chiefs of Staff exercises.

5-63. The joint combined exchange training program deploys SF teams overseas to train with foreign security forces. The typical joint combined exchange training is executed by an ODA; however, multiple ODAs training under the direction of an ODB are not uncommon. Additionally, ARSOF enablers, such as MISO, CA, and ARSOF aviation, may also participate to increase interoperability as well as train their mission-essential task lists. These organizations generally train with the HN/partner-nation forces in doctrinal disposition, generally training 30 to 100 HN participants per ODA. The joint combined exchange training program is a series of joint and combined SOF deployments for bilateral training conducted in an area of responsibility.

5-64. Joint combined exchange training events cover a wide range of activities. The primary focus of the program is to improve unit and individual combat readiness of U.S. SOF while providing an ancillary benefit to the friendly armed forces and enhancing bilateral relations and interoperability throughout the region. (Military-to-military contacts are covered under Title 10, U.S. Code, Section 168, *Military-to-Military Contacts and Comparable Activities*.) Secondarily, the joint combined exchange training program directly supports regional stability throughout the theater. Training typically focuses on small-unit tactics, marksmanship, medical skills, command and control, and other FID, counterinsurgency, and UW training mission-essential tasks. These have specific authorities under the following:

- Title 10, U.S. Code, Section 401, Humanitarian and Civic Assistance Provided in Conjunction With Military Operations.
- Title 10, U.S. Code, Section 402, Transportation of Humanitarian Relief Supplies to Foreign Countries.
- Title 10, U.S. Code, Section 2561, Humanitarian Assistance.

5-65. In 1991, Title 10, U.S. Code, Section 2011, *Special Operations Forces: Training With Friendly Foreign Forces*, codified the "special operations forces exception" to use Title 10, U.S. Code, funds to train with armed forces or other security forces of a friendly nation. This statute origin was based on a 1986 Comptroller General opinion recognizing a SOF requirement to train foreign forces based on U.S. SOF FID and UW missions. Based on the culmination of several years of supported legislation by the DOD, the Assistant Secretary of Defense for Special Operations and Low-Intensity Conflict, and the USSOCOM, Section 2011 authorizes funding expenses directly related to deploying and training U.S. SOF tasked with training friendly foreign forces overseas. Additionally, specific incremental expenses are authorized for funding when training with nations listed on the Secretary of State's Developing Country List. USSOCOM Directive 350-3, *Joint Combined Exchange Training*, provides additional guidance on the conduct of joint combined exchange training events. This authority is separate and distinct from that contained in the Foreign Assistance Act. Training evolutions conducted under Section 2011 are called joint combined exchange training events. Joint combined exchange training is an overseas training event for which the primary purpose must be to train U.S. SOF. Additionally, the training events must use Section 2011 reporting procedures (commonly referred to as the "2011 report"), must support and be prioritized through the GCC's theater campaign plan, and are primarily Major Force Program-11 funded. Finally, joint combined exchange training events are prohibited from using security assistance/security force assistance funds, are coordinated and approved by the U.S. Ambassador and Department of State, and approved by the Secretary of Defense.

5-66. The term joint combined exchange training cannot be used in conjunction with security assistance/security force assistance, counterdrug, or humanitarian assistance events. SF can conduct training with both conventional forces and civilian government agencies. Some joint combined exchange training events have a particular focus—such as medical, air operations, airborne operations, and maritime skills—in which SF has the opportunity to train on specific or critical mission-essential task lists.

5-67. U.S. SF teams are ideally suited for these types of engagements. They are highly trained, flexible, and adaptable. They can adjust quickly to meet the needs of the partner nation with whom they are training. In addition, SF units are regionally oriented, with personnel spending most of their military career with a specified group; for example, 5th SFG operates in the Middle East region, while the 1st SFG regional orientation is Asia. Thus, SF Soldiers not only participate in the HN's training activities, they have learned to respect the customs, speak the language, and often participate in HN special cultural activities and functions.

5-68. The joint combined exchange training program provides excellent combined training opportunities for all nations involved because it emphasizes the skills required to meet the most likely future military scenarios at the lower end of the range of military operations. These exchanges also enhance interoperability among the militaries of the region and promote regional stability. Thus, the joint combined exchange training program serves many vital needs in multiple regions.

COUNTERNARCOTERRORISM TRAINING

5-69. Counternarcoterrorism training utilizes U.S. SOF in conjunction with law enforcement agencies to provide country team-requested training to partner-nation law enforcement agencies and military units that have a counterdrug mission. Legislation enacted in 1989 tasked the DOD to serve as lead agency in detecting and monitoring illegal drugs entering the United States. The Armed Forces were also authorized to provide counterdrug-related training and support to domestic and foreign law enforcement agencies. The President outlined a comprehensive strategy for reducing illegal drug use in the United States and assisting friendly nations in combating both consumption and trafficking. The counternarcoterrorism program focuses on demand reduction and drug-related crimes.

5-70. The purpose of the counternarcoterrorism program is to enable (through training) foreign police and military to act as a force multiplier in the fight against drug flow into the United States and assist partner nations in controlling illicit activities in littoral waters and on land. All training is requested by the American Embassy and usually sponsored by a U.S. law enforcement agency or a non-DOD entity mandated by the Deputy Assistant Secretary of Defense for Counter Narcotics and Global Terror. Training topics include human rights; tactical movement; marksmanship (basic/advanced); patrolling; mission planning (police format); tactical medicine/trauma; arrest procedures; small boat operations; visit, board, search, and seizure (compliant/non-compliant); close-quarters battle (police format); and investigative procedures. Counternarcoterrorism is an excellent joint, interagency, intergovernmental, and multinational program that supports the theater security cooperation plan for the theater. Engagement with foreign police and the military provides access and placement in regions that no other program achieves.

5-71. Counternarcoterrorism utilizes multiple funding codes for execution. For example, the Joint Interagency Task Force-West counternarcoterrorism program utilizes Project Code 6504 directly from the Deputy Assistant Secretary of Defense for Counter Narcotics and Global Terror through USSOCOM for SOF-supported counternarcoterrorism. Project Code 6504 funds are budgeted by the Joint Interagency Task Force-West and can only be used for forces subordinate to USSOCOM and only when authorized by Joint Interagency Task Force-West. Project Code 9202 funds are used to pay for law enforcement agency integration into counternarcoterrorism. Support includes maintenance, repair, or upgrading of equipment; transportation of personnel and their equipment; the establishment of bases of operations or training facilities for counterdrug activities; associated support expenses for trainees; establishment of command and control networks; linguist services; and aerial reconnaissance. The National Defense Authorization

Act, Sections 1021 and 1022, authorizes support and names the countries eligible to receive support for counterdrug activities. Counternarcoterrorism is assistance provided to select countries to combat the influence of the international criminal drug enterprise on the security and stability of sovereign nations. Detachments conducting this training are authorized to train, advise, and assist HN military and police elements to develop effective forces capable of countering the narcoterrorist threat.

EXERCISE PROGRAMS

5-72. Exercise programs are designed to test ARSOF capabilities to operate in realistic, stressful, joint, and combined arms training according to joint, Army, and ARSOF doctrine. Combat Training Center programs are events in the continental United States that integrate SOF with conventional forces and conduct missions that are relevant to the unit mission-essential task list and the area of responsibility. Mission command and staff training ensures currency in procedures for conventional forces interoperability and operational synchronization and communications system technology. Programs outside the continental United States are mission-essential task list-related training and support the GCCs and U.S. Ambassadors with relevant training and regional expertise.

THEATER JOINT CHIEFS OF STAFF EXERCISE PROGRAM

5-73. The Theater Joint Chiefs of Staff Exercise Program provides SF units with opportunities to conduct mission-essential task list-related training while participating in joint and combined training events in HN countries. Joint Chiefs of Staff/GCC exercise programs are demonstrations and assessments of ARSOF mission capabilities. Joint Chiefs of Staff exercises are developed by the GCCs and TSOCs and coordinated with USSOCOM during USSOCOM's semiannual Global Synchronization Conference. Joint Chiefs of Staff exercises are included in the USSOCOM Global Force Management Allocation Plan, the database that lists all participating units, locations, dates, and cost estimates. Modifications to the Global Force Management Allocation Plan are published by USSOCOM as changes occur. Joint Chiefs of Staff exercises are funded using Joint Chiefs of Staff exercise funds, which are distributed annually by USSOCOM to USASOC for disbursement to the USASFC and the SFGs.

UNITED STATES SPECIAL OPERATIONS COMMAND JOINT NATIONAL TRAINING CAPABILITY EXERCISE PROGRAM

5-74. The purpose of the continental United States Joint National Training Capability exercise program is to provide SFG headquarters training in a joint environment. Generally, these exercises are sponsored by the USSOCOM as the executive agent of DOD and Joint Chiefs of Staff to conduct joint SOF training. USSOCOM sponsors field training/mission rehearsal exercises, which are accredited exercise programs in support of Chairman of the Joint Chiefs of Staff Instruction 3500.01G, *Joint Training Policy and Guidance for the Armed Forces of the United States*. Headquarters, USSOCOM, directs U.S. SOF, conventional forces, and interagency participants through realistic combat training in an irregular warfare scenario. This training is conducted in an integrated tactical and operational joint national training capability exercise environment, leveraging previous combat operations, previous UW exercises, and other lessons learned. USASFC uses these exercises to conduct SOTF-level joint training and appropriate force modernization equipment testing.

Chapter 6

Planning Considerations for Unconventional Warfare and Foreign Internal Defense

The strategy is one to ten; the tactics are ten to one.
Li Tso-peng commentary on Mao's *On Protracted War*

SF operations are often indirect and of long duration by nature. Because of the SF regional orientation, knowledge, language ability, and detailed cultural understanding, they are ideal for sensitive strategic operations that require a minimal U.S. footprint. SF units understand, anticipate, and plan for second- and third-order effects of actions taken on the ground in HN countries. SF operations inherently combine joint and interagency operability and require extensive deliberate planning in order to be successful. Special warfare operations, such as UW and FID, have national- and strategic-level implications and require an intensive amount of planning to ensure success.

UNCONVENTIONAL WARFARE CONSIDERATIONS

6-1. Planning for UW is slightly different from planning for other special operations. Most SF principal tasks—other than UW, FID, and counterinsurgency—are typically short-term operations that are tactically and conceptually simple. UW, however, usually involves long-term campaigns of greater complexity. UW campaigns require operational art to put forces in space and time; to integrate ends, ways, and means from the tactical to the strategic levels; and to attain the desired strategic effects and U.S. political or military objectives. The sensitivity of the planned action dictates the level of compartmentalization the United States must use to ensure operational security. Parallel planning by all echelons helps ensure that each echelon understands how their mission integrates with the missions of others.

6-2. The seven phases of U.S.-sponsored UW is the classic conceptual template that planners use to aid understanding of a UW campaign effort. As a template, it merely serves as a guide for planning and execution. The seven phases are—

- Preparation.
- Initial contact.
- Infiltration.
- Organization.
- Buildup.
- Employment.
- Transition.

6-3. The seven-phase template is designed with SF as the primary actor in all phases. With the exception of MIS forces, no other SOF will take part in all seven phases, but they may support a specific portion or phase of the larger UW campaign. Regardless, operational personnel should understand how their efforts integrate with and contribute to the overall campaign plan. It is possible that, in some cases, SF may not conduct all phases unilaterally and may continue or integrate into efforts initiated by other agencies.

Chapter 6

INTELLIGENCE

6-4. The preparation phase begins with the President and/or Secretary of Defense approval to execute a UW campaign. The primary purpose of this phase is to ensure that the insurgency, resistance, and the population are "prepared" to conduct and support a UW campaign. Years of preparation of the environment may have set the conditions for quick transition from shaping to preparation upon approval. This phase will include both information operations targeting specific audiences to interagency efforts to determine goals, capabilities, and liabilities of the insurgency or resistance before making initial contact. If, during this phase, it is determined that the insurgency, resistance, or the population are not supportive of the effort, then the operation may be cancelled.

Political and Cultural Factors

6-5. The preparation phase must begin with a full area assessment of the operational environment. UW operations need to include, but are not limited to, a thorough analysis of the resistance force's strengths, weaknesses, logistics concerns, level of training and experience, political or military agendas, factional relationships, and external political ties. Along with this data, a thorough area study of the target area should be done. At a minimum, the area study should include government services and political, religious, economic, weather, standard of living, medical, and education issues.

6-6. The U.S. Government begins to shape the target environment as far in advance as possible. The shaping effort may include operations to increase the legitimacy of U.S. operations and the resistance movement, building internal and external support for the movement, and setting conditions for the introduction of U.S. forces into the UW operating area. MIS assessments are particularly important during the preparation phase because they provide the ODA with vital information on possible insurgent leaders and key communicators that have psychological relevance with the population. Personnel could conduct these activities proactively in areas under the control of adversarial regimes or reactively immediately following an act of aggression against an ally's territory, such as an invasion. The population of a recently occupied country may already be psychologically ready to accept U.S. sponsorship, particularly if the country was a U.S. ally before its occupation. In other cases, psychological preparation may require a protracted period before yielding any favorable results.

6-7. Before the U.S. Government decides to render support to a resistance, it establishes contact with representatives of a resistance organization to assess the compatibility of U.S. and resistance interests and objectives. This assessment is largely a political negotiation between the U.S. Government and the resistance organization. Once the U.S. Government establishes compatibility, it assesses the resistance potential. During the initial contact, planners may arrange for the reception of a pilot team. If possible, planners may exfiltrate a resistance representative (referred to as an asset) from the operational area to brief the pilot team during its planning phase and possibly to accompany the team during their infiltration into the operational area and linkup with resistance forces.

6-8. The pilot team conducts detailed area assessments to expand its understanding of the operational environment, particularly civil considerations analysis. This analysis provides information on the degree of support for the UW effort among the local populations. MIS planners provide the pilot team with area- and population-specific information requests to facilitate the evaluation of indigenous information capabilities and the determination of the level of support necessary to fully develop those capabilities and increase their operational effectiveness. A pilot team is an ODA that may be modified or augmented as needed to meet specific mission requirements.

6-9. It is important that the ODAs identify the strengths and weaknesses of the resistance group and communicate these to higher headquarters so that higher headquarters can develop the UW campaign plan to effectively leverage the strengths of the different groups while mitigating the inherent weaknesses. The utilization of MIS assessments provides critical sources of information that aid in understanding these strengths and weaknesses. These assessments assist in comprehending the cultural, religious, economic, and social factors affecting the operational environment and the resistance movement. This analysis also provides key insights into relationships and other influences affecting the behavior of targeted groups.

6-10. The first step toward determining appropriate targets, objectives, and other supporting efforts is to establish an understanding of the enemy's capabilities and intentions. Key questions include the following:
- What are the enemy's tactical, operational, and strategic capabilities and efforts to ensure control over the population?
- What are the enemy's centers of gravity?
- What are the enemy's psychological vulnerabilities?
- Where is the enemy vulnerable to guerrilla or underground operations?

6-11. Once U.S. advisors have an accurate assessment of the situation on the ground, a good working relationship, and a concept for expanded operations, resistance leaders and planners work out the specifics of the concept. The two parties work out these specifics at all levels, clearing them through the U.S. interagency to ensure concept agreement. Depending on the length of the campaign effort, it is highly probable that the theater campaign plan was still in development at the time of the ODA isolation planning and infiltration. The intelligence derived from the resistance in this phase is critical to the theater planning efforts.

Balancing Risk to Mission Vice Risk to Force

6-12. Successful planners weigh the benefits of providing support to resistance forces against the overall strategic framework of a campaign. They must not allow a desire to conduct UW or to produce a purely military effect to dominate their judgment. Support to resistance forces does not simply contribute to a military effort; it undoubtedly alters the geopolitical landscape of a given region. Planners may deem a specific insurgent effort feasible and appropriate to the military effort, but consider it strategically unfavorable because of the political risk of the effort or the potential for increased regional instability.

6-13. Planning, therefore, remains limited until leadership validates certain assumptions. If operations proceed without a proper feasibility assessment, the likelihood of unintended consequences is high. To gain an accurate picture, operational personnel need to meet with indigenous personnel who represent the resistance forces. This meeting can take place inside the denied territory, in the United States, or in a third-party nation. Although meeting representatives in the United States or a third-party nation is safer for an assessment team, it also provides a less reliable assessment of potential capabilities. Participation of all components is vital to enable an accurate assessment of potential resistance capabilities.

6-14. The feasibility assessment is an assessment based on mission, enemy, terrain and weather, troops and support available, time available, and civil considerations to determine if the necessary means and resources are available to meet mission requirements. It analyzes the feasibility, adequacy, and acceptability of a mission and addresses whether the potential gain or desired effect outweighs or otherwise justifies the potential losses or cost.

Mission Command

6-15. Military commanders must understand that the mission command tactics and techniques used in other special operations do not transfer well to UW missions. Units cannot communicate with their headquarters in the same manner as during other types of operations. Even if the communication architecture is available, leaders must exercise great care before placing requirements on units operating from within enemy territory.

6-16. Unlike conventional units, UW organizations risk some degree of exposure with every communication. Personnel must not confuse communications encryption with low electronic signature. Leaders must balance a headquarters commander's desire for real-time communications with the constraints of the operational environment. Units engaging in UW must operate in a decentralized manner. They should always operate under the assumption that the enemy is trying to locate their position using unusual signals in urban and rural areas.

6-17. Unlike conventional operations, the acceptable size and optimum location for the unit headquarters engaging in UW changes as the mission progresses. The headquarters should base its decision to execute mission command from an adjacent country or from an infiltrated point in the resistance area where it can provide the most value added. Before entering territory under the control of the resistance, the unit headquarters must consider if their signature will jeopardize the mission and if the resistance headquarters could benefit from their direct interaction. In either case, a large headquarters footprint is inappropriate for these types of operations, particularly in forward areas.

6-18. The U.S. Army specifically designed the SFG to operate in a decentralized manner, synchronizing the efforts between displaced sector and area commands with their ODAs and ODBs. The group provides ODAs, SF companies operating as AOBs, and SF battalions operating as SOTFs to perform tactical functions in addition to their inherent mission command responsibilities. Each one of these headquarters maintains the ability to operate with its equivalent resistance-force counterpart.

Relationship With Resistance Forces (Influence Versus Command)

6-19. Once U.S. advisors link up with resistance leadership, the objective is to determine and agree upon a plan to organize the resistance for expanded operations. In addition to physical preparations, this plan entails a confirmation of mutual objectives and prior agreements. This requires a period of rapport building to develop trust and confidence, as well as a period of discussion of expectations from both sides.

6-20. Many guerrilla leaders may not enthusiastically accept U.S. advisors but may tolerate them as a precondition for other U.S. support, such as logistical aid. They may harbor suspicions that the U.S. intentions are purely self-serving or lack the resolve to maintain their commitment long term. Guerrilla leaders are cautious of placing too much reliance in U.S. promises. It is the challenge of the ODA leadership to gain the confidence of the resistance leadership and demonstrate the value of cooperation toward their mutual goals. The ODA explains its capabilities and limitations and begins to assist the resistance leadership with the development of the organization. Although rapport eases every aspect of operations, operational personnel must not perceive rapport as the goal, but rather a means to a goal. The goal is a strong relationship in which the SF leaders are trusted advisors who can influence the direction of events. The amount of influence an advisor attains is directly proportional to the total sum of three factors: rapport, credibility, and the continued belief in the value of the relationship.

6-21. The resistance leader and U.S. advisors must agree upon command and control arrangements within the bounds of higher-level political and military agreements. The specifics of a resistance organization depend on local conditions. UW requires centralized direction and decentralized execution under conditions that place great demands on the resistance organization and its leadership. Armed rebellion inherently creates an ambiguous and unstructured environment. No two resistance organizations need the same degree or level of organization.

6-22. Before a resistance organization can successfully engage in combat operations, its leadership must organize an infrastructure that can sustain itself in combat and withstand the anticipated hostile reaction to armed resistance. During the organization phase, the resistance leadership develops a cadre to serve as the organizational nucleus during the buildup phase. The ODA assists the resistance leadership in this endeavor by providing training and advice to prepare for the eventual buildup of the resistance organization.

Unified Action With Interagency

6-23. Because of the military and political nature of UW, the U.S. interagency involvement is critical to achieving a holistic government approach and long-term success. The full integration of joint, interagency, intergovernmental, and multinational communities is necessary at various stages of an unconventional conflict. The United States conducts two types of UW—with or without the anticipation of large-scale U.S. military involvement. The extent of interagency involvement will differ depending on the type of UW being prosecuted.

6-24. During large-scale U.S. military involvement, UW, operations focus largely on military aspects of the conflict because of the eventual introduction of conventional forces. The task is normally to disrupt or degrade enemy military capabilities in order to make them more vulnerable to the pending introduction of conventional invasion forces. The interagency plays a supporting role to the military instrument of power when large-scale military operations are planned. This role will vary with the degree of declared U.S. contention with the target state, and will consist largely of political influence with surrounding states and with international opinion.

6-25. During limited-involvement missions, the overall operation takes place in the absence of overt or eventual hostilities from the sponsor. Such operations take on a strategic aspect of heightened political sensitivity.

6-26. Typically, the United States limits its direct involvement, which mitigates the risks of unintended consequences or premature escalation of the conflict. During limited-involvement operations, the manner in which forces operate significantly differs from that of large-scale involvement scenarios. Without the benefit of a conventional invasion force, the resistance forces must limit overt exposure of their forces and supporting infrastructure in order to sustain operations over a protracted period. Forces must conduct operations in a manner that accounts for the enemy's response and retaliation. In the absence of overt hostilities, the interagency plays a leading role typical of normal peacetime engagement. In such circumstances, supporting interagency activities will emphasize discrete and indirect efforts to shape international opinion, and the mustering of political support and logistical aid for the insurgent causes. Moreover, interagency access and intelligence assets may support the clandestine military effort.

MOVEMENT AND MANEUVER

6-27. Successful infiltration of ODAs and right-sized mission command ensures theater or countrywide unity of effort and apportionment of resources. This represents a decisive point for the operation, as well as a period of increased operational risk. Infiltration plans and tactics need to remain focused on successfully achieving this decisive point.

Emphasis on Low-Visibility Methods and Platforms

6-28. During the infiltration phase, a pilot team clandestinely infiltrates the operational area in order to link up with the resistance force and conduct or confirm a feasibility assessment. If the operational assessment is favorable, the pilot team coordinates for the infiltration and reception of follow-on SF teams and supplies. Depending on the situation, circumstances may dictate the infiltration of ODAs without the benefit of a prior pilot-team effort, a trusted asset, or a completed feasibility assessment. In this case, the ODA will perform many of the required functions normally accomplished by the pilot team. If this occurs, the ODAs may need to adjust their infiltration plans to account for the higher risk of infiltration without the benefit of a reception coordinated by other U.S. personnel.

6-29. In limited-war scenarios where the infiltration of U.S. personnel is undesirable, planners could exfiltrate indigenous resistance personnel out of the target area, provide training in specific required skills, and infiltrate the personnel back into the target area to function as cadre capable of conducting operations or training other resistance forces. The enemy's level of control over the population and the environment affects how long planners can keep resistance personnel away from their region before they begin to disconnect from the local population.

Clandestine Planning Considerations

6-30. Planners cannot fix the organization of the guerrilla force according to standard, conventional tables of organization and equipment. Guerrilla force missions and tactics dictate a simple, mobile, and flexible organization capable of rapid dispersion and consolidation. Each unit must function as an independent organization with its own intelligence, communications, and logistical systems. Guerrilla organization normally determines auxiliary organization. Planners should compartmentalize all auxiliary functions from one another and from the guerrilla force that the auxiliary supports. The underground will require the most discretion of all resistance elements, as they typically operate in urban or built-up areas under tight enemy control and close observation. Undergrounds use a range of techniques to ensure their activities remain compartmented, redundant, and both covert and clandestine.

FIRES

6-31. Supporting fires, such as air strikes used by SF advisors against the Taliban in Afghanistan in 2001, are used occasionally. Other technologies, such as naval surface fire support or missile-equipped unmanned aircraft systems, can also be employed. Moreover, an insurgency may reach a level of organization and strategic success where they can field their own artillery assets, as did the Chinese during the Anti-Japanese and Chinese Civil Wars of the 1940s. However, because of the typical need for discretion by external supporters—especially in early high-risk phases—the majority of fires in support of UW will be nonlethal.

MILITARY INFORMATION SUPPORT TO UNCONVENTIONAL WARFARE

6-32. MISO exploit resistance or insurgency successes for their maximum psychological effect. This exploitation can increase the morale of resistance forces and auxiliaries, which can further increase their operational effectiveness. In addition, success brings positive attention to the movement and increases support from the indigenous population and external supporters in terms of logistics, intelligence, and recruitment. MISO also exploit resistance or insurgency successes to erode enemy morale and decrease internal and external support. MISO may also increase dissension, desertion, and surrender of enemy forces, further decreasing their operational effectiveness. MISO can further exploit enemy reprisals against populations or the guerrillas to separate the population from the enemy government or occupying forces.

U.S. and Indigenous Assets

6-33. Physical attacks that resistance forces conduct can significantly alter psychological effects during this phase. Planning actions for psychological effect is a deliberate process requiring thorough analysis, detailed coordination, and careful execution. Although this process is time- and labor-intensive, the effects can potentially shape the course of the entire UW operation in a profound manner. SF units on the ground coordinate and synchronize these efforts to ensure a complementary effect, first with partner force and then conventional force efforts. Other supporting efforts may include—

- Gathering and reporting vital intelligence to coalition forces.
- Assisting in the evasion and recovery of isolated personnel (downed aircrews).
- Reconnoitering and receiving airborne, air assault, or amphibious invasion forces.

6-34. MISO can enhance the effects of these supporting efforts. They can increase actionable intelligence obtained from key segments of the population through persuasive messages that increase sympathy and support for the resistance movement. Information on rewards and other messages can persuade target groups to aid the evasion and recovery of isolated personnel. In addition, MISO can assist in building local networks that provide support for incoming invasion forces by consistently emphasizing the benefits of supporting the UW effort and highlighting the negative aspects of the enemy government or occupying power. Finally, MIS elements can act as force multipliers by training indigenous assets to integrate information-related activities into operations to influence populations.

Nonlethal Influence on Target Population

6-35. UW operations have a fundamental psychological component in both execution and effects. Consequently, UW planning should integrate MIS forces in all phases of the operation. MISO can develop, maintain, and reinforce desired behaviors in target groups and individuals, while minimizing undesired behaviors. MISO are the primary method of fulfilling the ARSOF imperative of anticipating and controlling psychological effects. A key task of controlling psychological effects is ensuring that populations in the UW operating area understand that operations are for their ultimate benefit, even if not immediately so. Failure to shape popular perceptions in ways that support UW objectives leaves the operation vulnerable to adversary information activities that can negatively affect even the best-planned and executed missions.

PROTECTION

6-36. UW operations present some unique personnel protection challenges. Because of the low signature of many UW operations, the normal measures for protection may be impractical or impossible. Personnel can mitigate some aspects of the associated risk through good operational security measures and signature reduction measures.

6-37. During the buildup phase, the resistance cadre improves the organization's clandestine supporting infrastructure to prepare for expanded offensive operations. When the organization begins to conduct operations of a wider scope and across a wider area, many of these operations will draw attention from counterguerrilla forces. The organization must have the supporting clandestine infrastructure to prepare for and sustain these operations.

OPERATIONS SECURITY

6-38. There are numerous aspects of operations beyond the control of advisors. As such, advisors should always employ proactive measures to protect operational information regardless of their estimate of the actual risk at hand. Compromises of operational information may occur hundreds of miles from the operational area. Personnel should take the following precautions:

- Limit the use of proper names with resistance members. Soldiers should not share personal information with indigenous resistance personnel.
- Provide code names for all advisors. This allows secure and unsecure communication regarding ODA personnel.
- Keep operational information on a need-to-know basis.
- Maintain internal communications procedures that indicate a compromise of information.

6-39. The situation may dictate that U.S. personnel reduce their distinctive appearance. This may include the wear of articles of indigenous clothing, such as a scarf or hat; the carrying of indigenous weapons and gear; or the adoption of normal customs of appearance, such as a mustache or beard. Signature reduction measures can allow SF personnel to blend in with indigenous forces and prevent identification by enemy personnel. The unit chain of command must approve any deviations from standard appearance and must ensure they comply with the legal requirements of the operation.

6-40. Auxiliary contributions to security are also important. The auxiliary refers to that portion of the population that provides active clandestine support to the guerrilla force or the underground. Members of the auxiliary are part-time volunteers that have value because of their normal position in the community. Soldiers should not think of the auxiliary as a stand-alone organization. They provide specific functions supporting urban underground networks and guerrilla forces. Because auxiliary members conduct their daily civilian activities out in the open, they have more ability to travel wherever they are needed than the other components of the resistance. Therefore, the auxiliary is essential for providing outer cordons of security and early warning for underground facilities and guerrilla bases.

SUSTAINMENT

6-41. Logistics support for UW is different from support to other types of special operations and is even different from most other instances of special warfare. UW missions often require significant quantities of materiel to support resistance forces, specifically guerrillas. The materiel includes lethal and nonlethal aid, some of which may not be organic to the U.S. military supply system. Every effort must be made to maximize the use of indigenous supply sources within the UW operating area. In addition, confiscation, barters or trades, "IOUs," donations or levies, and battlefield recovery and purchase is leveraged extensively in order to maximize demands of external resupply. Planners need to consider the ability of the indigenous forces to make use of U.S. materiel. In some cases, introduction of some U.S. materiel may not be compatible with other materiel found in the operational environment and may complicate resistance forces' efforts.

6-42. The lack of established lines of communications presents another significant challenge. Personnel must deliver materiel in a manner that does not compromise the indigenous force's location. For this reason, planners need to carefully consider and prioritize external resupply efforts, keeping resupply to essential items only. Planners need to ensure resupply efforts do not establish a pattern. Personnel should use various methods and locations throughout the course of the buildup and combat employment phases.

Delivery of Sustainment in Denied Areas

6-43. Once planners determine the type and scope of operations, they develop supporting capabilities specific to those efforts. Tasks outlined in the UW campaign plan or operations order drive supporting capabilities. The ODA may start to coordinate for specific supplies via airdrop or other resupply methods. Planners need to prioritize resupply efforts for materials that forces cannot procure by other means. Every resupply operation comes with the risk of exposure and potential loss of the supporting apparatus. Planners need to develop and emplace capabilities without compromising the organization or future operations. If resistance efforts are to support a pending coalition D-Day (start of the coalition invasion), the capabilities—whether guerrilla, auxiliary, or underground—need to include notification and activation procedures that allow the synchronization of efforts with conventional forces.

6-44. There are two main categories of resupply—accompanying resupply and external resupply. External resupply comprises of automatic, emergency, and on-call (or routine) resupply. The ODA may take accompanying supplies into the joint special operations area at the time of infiltration. The threat in the joint special operations area dictates the quantity and type of supplies and equipment the ODA can include. External resupplies are procured and delivered to the UW operating area/joint special operations area by the sponsor (JSOTF), based on the needs of the resistance force or insurgents, as well as the ODA. Resupply is planned in isolation to be delivered after infiltration at a coordinated location and time automatically, as requested, or based upon a no-communications trigger.

6-45. The purpose of the emergency resupply is to provide essential equipment and supplies to restore operational capability and survivability of the ODA. Typical items contained in the bundle may be communications equipment, batteries, weapons, ammunition, money, and handheld global positioning systems. A coded message, a radio request, or the absence of any detachment communication over a prearranged period can trigger an emergency resupply. The ODA and the supporting headquarters must clearly understand the sequence of events, time required, and assets available to deliver the emergency resupply.

6-46. When the ODA establishes communications with the JSOTF or SOTF, external supply begins on call. Personnel use the abbreviated code of a catalog supply system contained in the signal operating instructions to request supplies based on operational need. These supplies consist of major equipment items that units do not consume at a predictable rate. Theater Army area command depots, the JSOTF, or the SOTF hold these items in readiness for immediate delivery on a specific mission-request basis.

6-47. The preferred mission delivery method for external resupply is by sponsor aircraft, surface ship, or submarine. At first, planners may determine aerial delivery by parachute is the best means of supply to a joint special operations area. Personnel may use free-drop techniques for certain hardy items. Later, as the joint special operations area expands and comes under greater friendly control, ODA members use air-landed supply missions. Supply personnel normally use surface ships or submarines when joint special operations areas are next to waterways or seas. Resupply operations require secrecy to protect the resupply platform and the reception element. Personnel normally conduct these operations during limited visibility.

Indigenous Sources of Sustainment

6-48. Each resistance organization must develop a logistics system to meet the specific requirements of their situation. In general, however, a resistance organization meets its logistical requirements through a combination of internal and external means. The area complex must provide the bulk of an insurgent organization's logistical requirements. The area commander must balance the support requirements against the need for civilian cooperation. Imposing excessive demands on the population may adversely affect popular support. Logistical constraints may initially dictate the size of the resistance organization.

6-49. As the resistance organization expands, its logistical requirements may exceed the capability of the area complex to provide adequate support. When this situation occurs, an external sponsor provides supplemental logistical support or the resistance organization reduces the scale of its activities. External support elements normally limit support to the necessities of life and the essential equipment and supplies the resistance needs to conduct combat operations. Internal sources of resistance supply include the following:

- Battlefield recovery.
- Purchase.
- Levy.
- Barter.
- Production.
- Confiscation.
- Procurement of nonstandard items.

6-50. Nonstandard items describe items that are not stocked, type-classified, or safety-approved for use by U.S. forces. The acquisition and infiltration of nonstandard items into the area of operations can be very important to help maintain low visibility of indigenous resistance forces where foreign equipment would otherwise draw unwelcomed attention. Military and interagency organizations have developed contact with international providers for many items useful to the resistance organization.

LEGAL CONSIDERATIONS

6-51. SF conducts UW operations in accordance with U.S. domestic and international law. The baseline U.S. legal authority for UW is in Title 10, U.S. Code, Section 167(j), *Special Operations Activities*. This provision states that UW is one of the core activities of USSOCOM. Per Title 10, USSOCOM has the authority to prepare, train, equip, fund, and sustain forces for UW. Personnel can only conduct actual UW operations with specific approval and authorizations from the President or Secretary of Defense. The foremost legal concern for UW operations is that every proposed operation receives a specific legal review and that all aspects of planning and executing UW operations are closely coordinated with legal advisors.

6-52. UW operations involve many unusual and often unsettled legal matters, including authority to conduct operations, funding, legal status of personnel, and a host of other issues. The legal parameters of UW are rarely clear and depend on the specifics of a particular mission, campaign, or conflict. SF should know the potential that individual and small-unit UW operations have to affect matters on the international level. SF must possess awareness of the standards that apply to UW and the implications of conducting UW under U.S. and international law. Because of its nature, UW requires close coordination with legal advisors in all phases of planning and executing operations of this type.

6-53. Law and policy in the area of UW are subject to rapid change. New U.S. congressional legislation and Presidential executive policy can affect and often directly address UW operations. New treaties, United Nations actions, and shifting views among nations can dramatically change the international implications of UW operations. Consequently, SF commanders must draw on the expertise of their legal staff and the organic sociopolitical expertise of SF Soldiers and relevant subject-matter experts before establishing UW plans and policy. Constant monitoring of the legal ramifications of UW operations is also necessary.

6-54. There is no special body of law for UW. The usual operational authorities govern UW by applying the specific facts of each operation on a case-by-case basis. A baseline consideration for U.S. UW operations is that SF cannot do anything through irregular forces that SF could not legally do alone. UW is not a means to circumvent FM 27-10 or U.S. and international law. U.S. forces must comply with FM 27-10 in all military operations regardless of the nature of those operations. Further, all U.S. forces have a duty to report any violations of FM 27-10 to their chain of command, whether committed by U.S. forces, other regular forces, or irregular forces. (More detailed information on UW legal considerations can be found in TC 18-01.)

FOREIGN INTERNAL DEFENSE CONSIDERATIONS

6-55. SF operating overseas conducting a FID mission must understand the Department of State is the lead agency. JP 3-08 identifies the ambassador as the personal representative of the President to the government of a foreign nation. The ambassador is responsible for implementing national policy and as the senior U.S. Government official has extraordinary decisionmaking authority. The three dimensions of U.S. policy overseas generally encompass Department of State, U.S. Agency for International Development, and DOD efforts. Understanding the priorities and objectives of each of these stakeholders creates a basis for integrating FID efforts and achieving desired results.

DEPARTMENT OF STATE GUIDANCE AND PLANNING RESOURCES

6-56. Guidance is the first step in any planning process. Every four years the Department of State produces the Quadrennial Diplomacy and Development Review. This document is based on the President's National Security Strategy and is comparable to the Department of Defense Quadrennial Defense Review. The Quadrennial Diplomacy and Development Review directs U.S. civilian agencies to advance national interests abroad and guides the State Department's Joint Strategic Plan and each of its bureaus' plans. The six regional bureaus of the State Department are analogous to geographic combatant commands; however, they are not aligned in an identical geographical manner. Every three years the regional bureaus of the State Department produce a joint regional strategy document (which replaced the Bureau Strategic and Resource Plan) that provides priorities, goals, and strategic focus. Each ambassador develops his or her own integrated country strategy. An annex to that document is the mission resource request, which seeks the resources to implement the strategy. All of these documents inform the interagency security cooperation process. Reviewing these documents and recognizing the planning horizons aid the integration of military resources into the overall strategy.

DEPARTMENT OF DEFENSE GUIDANCE AND PLANNING RESOURCES

6-57. FID programs are a part of security cooperation, which are all DOD interactions with foreign defense relationships and provide U.S. forces with peacetime and contingency access to the host nation. JP 3-22, FM 3-22, and FM 3-05.2 provide a more complete discussion of security cooperation and FID. For SF planners engaged in FID programs, key military documents providing guidance are the theater campaign plan and the country campaign plan.

6-58. The theater campaign plan translates national strategy into operational concepts designed to achieve regional end states. This plan is the comprehensive integration of steady-state military regional activities (security cooperation and other shaping or steady-state activities). It also supports global campaigns such as counterterrorism and counterproliferation.

6-59. Sometimes published as an annex to the theater plan, the country campaign plan is where all guidance comes together in an actionable plan. The country-specific plan is usually the responsibility of the desk officer in the plans section at the combatant command. This plan is formed in collaboration with the security cooperation office, the country team, and the service component commands that provide the forces. Key elements in a country plan are objectives, activities, and resource requests.

6-60. Although higher guidance is integral in planning for FID, SF planners must understand that partner nations are sovereign and have their own strategies, capabilities, and perceived threats. U.S. influence on another country is always limited by that nation's desires and needs. FID plans are most successful where the interests of all parties coincide.

LEGAL CONSIDERATIONS

6-61. For U.S. Armed Forces to conduct operations a legal basis must exist. Orders, laws, policies, regulations, treaties, and directives influence all aspects of operations. Without a deployment or execution order from the President, U.S. Armed Forces can only make limited contributions to operations that involve FID. If the Secretary of State requests—and the Secretary of Defense approves—SF can participate in FID under various programs of Title 22 and Title 10, U.S Code. Each of these programs has its own specific authorities, appropriations, and limitations for foreign assistance and national defense. SF planners must consult with the supporting staff judge advocate to gain a full understanding of the legal underpinnings for operations. A more thorough treatment of FID legal considerations is provided in JP 3-22 and FM 3-22.

INTELLIGENCE OPERATIONS

6-62. Intelligence cooperation is enabled by an information-sharing environment that fully integrates joint, multinational, and interagency partners in a collaborative enterprise. U.S. intelligence cooperation ranges from strategic analysis to current intelligence summaries and situation reporting for tactical operations. Intelligence collection and dissemination capabilities are often weak links in the HN military capability. U.S. military communications hardware and operators may also be supplied in cases where HN infrastructure cannot support intelligence operations. The release of classified information to the HN is governed by national disclosure policy. Detailed written guidance may be supplemented with limited delegation of authority where appropriate.

6-63. The primary duty of intelligence personnel engaged in FID is to produce intelligence to prevent or defeat lawlessness or insurgency. The SF unit must be ready to train, advise, and assist HN personnel in intelligence operations. Intelligence personnel must collect information and produce intelligence on almost all aspects of the FID environment. When they know that insurgents, terrorists, or common criminals receive aid from an external power, intelligence personnel seek information on the external power's role in the insurgency. They need information not only on the armed insurgents but also on their infrastructure organizations and their relationships with the populace. These relationships make the populace a most lucrative source of information.

Collection Program and Insurgent Activity

6-64. A sound collection program and proper use of the various collection agencies and information sources will result in a very heavy volume of information flowing into the intelligence production element. Because of the insurgent environment, politics, and military tactics, intelligence personnel can meet intelligence requirements only by reporting minute details on a great variety of subject areas. Each detail may appear unrelated to others and insignificant by itself. However, these details, when mapped and chronologically recorded over long periods and analyzed with other reported details, may lead to definitive and predictable patterns of insurgent activity.

6-65. The insurgent recognizes the shortcomings in his military posture. Therefore, he must minimize the weaknesses inherent in using and supporting isolated, unsophisticated forces that use ponderous and primitive communications and logistics systems. He uses the weather, terrain, and populace, employing secrecy, surprise, and simplicity. Plans and actions these unsophisticated forces will carry out must be simple, comprehensive, and repetitive. Therefore, the solution to a problem is a system that as a whole is complex but in part is independent, having simple, logical, and uniform characteristics.

Intelligence Activities

6-66. Intelligence activities provide the military leadership with the information needed to accomplish missions and implement the National Military Strategy. Planners use intelligence to identify the threat's capabilities and centers of gravity, project probable courses of action, and assist in planning friendly force employment. The intelligence required is of the type, quantity, and quality that—
- Provides goals for daily or major operations (intelligence that locates guerrillas for tactical counterguerrilla operations).
- Enables HN forces to retain or regain the initiative.
- Enables HN forces to put continuous and increasing pressure on insurgent security.

Foreign Internal Defense Operations

6-67. Several areas deserve special attention when discussing employment of forces in FID operations, including public perception; psychological impact; intelligence support; SOF and/or conventional force selection; public information programs; logistic support; counterdrug operations; combating CBRNE operations; counterterrorism; operations security; and lessons learned. In addition, units assigned a FID mission must implement procedures to help the Department of State and the country team vet HN forces to ensure the identification of personnel with a history of human rights violations.

6-68. In FID operations, the targets are elements of the populace—either civilian supporters or members of the insurgency. The differences between supporters and members are usually ill-defined. A complete awareness and intimate knowledge of the environment is essential to conducting current intelligence operations. The basic nature of the internal security problem requires an intensive initial intelligence effort to pinpoint the roots of subversion.

UNDERSTANDING OF POLITICAL AND CULTURAL FACTORS

6-69. To assist a country with its internal defense and development efforts, the following factors must be understood: political climate, social attitudes, economic conditions, religious considerations, philosophy or plan of the insurgents, the host government, and the local population. In addition, it must be understood how the United States implements diplomatic, economic, informational, and military instruments in a coordinated and balanced combination to help remedy the situation.

6-70. Those governments that lack the will to address their social, economic, or political problems are unlikely to benefit from outside assistance. However, governments that do mobilize their human and material resources may find that outside help makes a critical difference. Where significant U.S. national interests are involved, the United States may provide economic and military assistance to supplement the efforts of such governments.

6-71. The creation of a relatively stable internal environment, one in which economic growth can occur and the people are able to determine their own form of government, is a primary U.S. objective. Economic assistance, either supplied by the United States through bilateral agreements or by several nations through multilateral agreements, may help achieve this objective.

6-72. The primary responsibility for creating a stable atmosphere through the commitment and use of all its internal resources rests with the threatened government. Under certain conditions, U.S. policy supports supplementing local efforts to maintain this order and stability. These conditions are as follows:
- The internal disorder is of such a nature as to pose a significant threat to U.S. national interests.
- The threatened country is capable of effectively using U.S. assistance.
- The threatened country requests U.S. assistance.

6-73. The U.S. Government spends billions of dollars a year, with certain expectations, in programs to improve multinational and friendly nations. There are numerous benefits for the U.S. military to conduct FID throughout the world. These benefits include the following:
- Favorable relationships that promote U.S. interests are built and fostered. In many cases, these programs lead to the establishment of personal and unit relationships.
- Friendly-nation capabilities are strengthened, which ultimately strengthen U.S. security concerns.
- Many of the foreign areas aided by the United States provide U.S. forces with peacetime and contingency access.
- The proficiency and skills of U.S. forces are increased through training exercises with foreign nations.
- U.S. forces' regional knowledge of specific areas is improved, which can be disseminated throughout the force (environment, terrain, social, political, economic, culture, and beliefs).
- Effectiveness of the global operations against terrorist networks is improved.

6-74. Subversion, lawlessness, and insurgency are the result of specific conditions within a nation. They may stem from the population's perception that they are suffering from such conditions as poverty, unemployment, religious disparity, political issues, crime, or tribal unrest. These conditions have historically set the stage for lawlessness and insurgent activity against an established government. This type of internal strife or conflict within a nation's borders may remain a local problem or expand, which allows an outside source to influence or create opposition toward the legitimate government. In some cases, outside sources may threaten the HN's stability by exploiting the conditions within that nation to further their own cause. This outside influence may even establish itself within the HN to promote and support civil unrest. These types of conditions promote insurgencies and their violent solutions, like terrorism. U.S. military involvement in FID has traditionally focused on support to the HN's counterinsurgency. Although much of the FID effort remains focused on this important area, U.S. FID programs may aim at other threats to the HN's internal stability, such as terrorism. Moreover, an increasing proportion of FID activities in the 21st century will focus on assisting partnered nations to build overall defense capacity.

6-75. Identification of the root cause of the problem, analysis of the environment, and identification of the specific needs of the HN are crucial in tailoring military support to assist the HN's internal defense and development program. Emphasis should be on helping the HN address the root cause of instability in a preventative manner rather than reacting to threats. The United States supports specific nations based on U.S. policy toward that nation or region and implements FID programs to support that nation through GCC security cooperation programs. FID programs of all types, such as professionalization training of partner security forces, humanitarian assistance, and counterterrorism programs can prevent, reduce, or stop mitigating factors that can contribute to the beginning or spread of terrorism and insurgencies. FID activities implemented through the GCC may ultimately lead to stability within that nation or region and effectively reduce threats to the United States.

MISSION COMMAND IN FOREIGN INTERNAL DEFENSE

6-76. The U.S. military can provide resources such as material, advisors, and trainers to support these FID operations. In instances where it is in the security interest of the United States, and, at the request of the HN, more direct forms of U.S. military support may be provided, to include combat forces. The following principles apply to FID:

- All U.S. agencies involved in FID must coordinate with one another to ensure that they are working toward a common objective and deriving optimum benefit from the limited resources applied to the effort. In almost all cases, the U.S. ambassador assigned to the HN is the supported key U.S. official.
- The U.S. military seeks to enhance the HN military and paramilitary forces' overall capability to perform their internal defense and development mission. An evaluation of the request and the demonstrated resolve of the HN government will determine the specific form and substance of U.S. assistance, as directed by the President.
- Specially trained, selected, and jointly staffed U.S. military survey teams, including intelligence personnel, may be made available. U.S. military units used in FID roles should be tailored to meet the conditions within the HN.
- U.S. military support to FID should focus on assisting HNs in anticipating, precluding, and countering threats or potential threats.

Command Relationships

6-77. TSOCs are of particular importance because of the significant role of SOF in FID operations. The TSOC normally has operational control of SOF in the theater and has primary responsibility to plan and execute special operations in support of FID. SOF assigned to a theater are under the combatant command (command authority) of the GCC. The GCC normally exercises this authority through the commander of the TSOC. When a GCC establishes and employs multiple JTFs and independent task forces, the commander of the TSOC may establish and employ multiple JSOTFs to manage SOF assets and accommodate task force special operations requirements. Accordingly, the GCC, as the common superior, will normally establish supporting or tactical control command relationships between JSOTF commanders and JTF/task force commanders. Coordination between the joint force special operations component commander (who is also the commander of the TSOC) and the other component commanders within the combatant command is essential for effective management of military operations in support of FID, including joint and multinational exercises, mobile training teams, integration of SOF with conventional forces, and other operations.

6-78. The GCCs integrate all military security assistance plans and activities with regional U.S. military plans. The role of the GCC is critical. Regional perspective for the commander is at the operational and strategic level of conflict. He identifies and applies military and certain humanitarian or civic action resources to achieve U.S. national strategic goals. With proper and timely employment, these resources minimize the likelihood of U.S. combat involvement.

6-79. If circumstances dictate, it may be necessary to expand U.S. assistance by introducing selected U.S. military forces. A JTF will normally be established to coordinate this effort. This JTF—

- Exercises operational control of assigned U.S. military forces.
- Plans and conducts joint and combined exercises in coordination with the armed forces of the host government.
- Executes area command responsibilities for U.S. forces to ensure unity of effort.
- Specifies the chain of command; however, units may be required to report to various organizations, to include Department of State.

6-80. Within the DOD, the principal element charged with providing advisory assistance is the security assistance organization. SF personnel may provide assistance in two ways: as an SF unit providing advice and assistance to the HN military or paramilitary organization or as an individual SF Soldier assigned or attached to the security assistance organization. In either case, SF may be under operational control of the security assistance organization chief in his or her role as the in-country U.S. defense representative. However, SF will usually be under operational control of a TSOC. The security assistance organization includes all DOD elements, regardless of actual title, assigned in foreign countries to manage security assistance programs administered by the DOD. The U.S. advisor may often work and coordinate with civilians of other U.S. country team agencies. When that occurs, the U.S. advisor must know their functions, responsibilities, and capabilities because many activities cross jurisdictional borders. The country team is composed of U.S. senior representatives of all U.S. Government agencies assigned to a country. Together, SF advisors and their counterpart must resolve problems by means appropriate to the HN, without violating U.S. laws and policies in the process. SF advisors operate under very specific rules of engagement with the purpose of ensuring that advisors remain advisors.

6-81. The SF advisor must understand the scope of security assistance organization activities. The advisor must know the functions, responsibilities, and capabilities of other U.S. agencies in the HN. Because many SF activities cross the jurisdictional boundaries or responsibilities of other country team members, SF advisors seek input from other country team members to coordinate their portion of the overall FID effort.

6-82. Even though the HN may refuse U.S. advisors, HN military leaders may request and receive other types of assistance, such as air or fire support. To coordinate this support and ensure its proper use, U.S. liaison teams accompany HN ground maneuver units receiving direct U.S. support. Language-qualified and area-oriented SF teams are especially suited for this mission.

6-83. Figure 6-1, page 6-15, shows one possibility for a mission command and advisory assistance relationship for a single AOB that is deployed to provide advisory assistance to a brigade-sized unit of the HN. In this instance, the AOB provides mission command systems and logistics for its subordinate ODAs, and advisory assistance to the brigade-level echelon. There are situations where the AOB may deploy and not provide advisory assistance while still providing mission command systems and logistics for its subordinate ODAs.

Note: By law, U.S. Ambassadors or chiefs of mission coordinate, direct, and supervise all U.S. Government activities and representatives posted in the foreign country to which they are accredited. Chiefs of mission do not, however, exercise control of U.S. personnel attached to and working for the head of a U.S. mission to an intergovernmental organization (for example, U.S. Ambassador to the North Atlantic Treaty Organization) or U.S. military personnel operating under the command of a GCC. Generally, each chief of mission has an agreement with the GCC delineating which DOD personnel fall under the responsibility of each for security (JP 3-08).

Planning Considerations for Unconventional Warfare and Foreign Internal Defense

Note: The senior defense official is the diplomatically accredited defense attaché. The senior defense official/diplomatically accredited defense attaché is the chief of mission's principal military advisor on defense and national security issues, the senior diplomatically accredited DOD military officer assigned to a U.S. diplomatic mission, and the single point of contact for all DOD matters involving the embassy or DOD element assigned to or working from the embassy. All DOD elements assigned or attached to or operating from U.S. Embassies are aligned under coordinating authority of the senior defense official/diplomatically accredited defense attaché (DOD Directive 5105.75, *Department of Defense Operations at U.S. Embassies*).

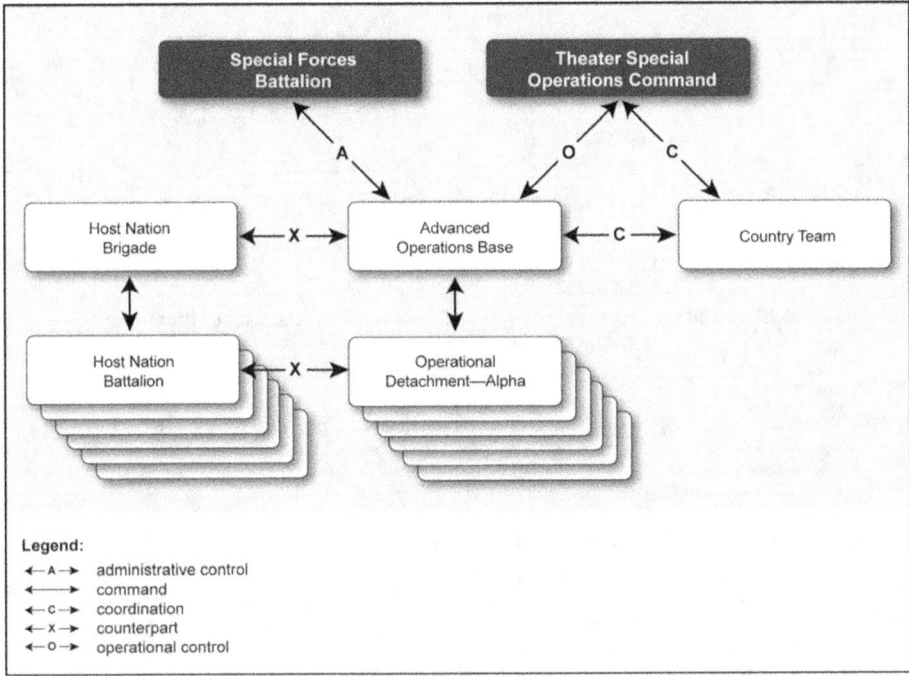

Figure 6-1. Advanced operations base providing mission command systems, logistics, and advisory assistance

6-84. Figure 6-2, page 6-16, shows a typical mission command and advisory assistance relationship for a single detachment deployed as part of the GCC/TSOC Theater Security Cooperation Plan. Such missions can include—but are not limited to—FID training on a broad range of subjects, counterterrorism, humanitarian assistance, and other missions to build and enhance partner capacity. Joint combined exchange training, Joint Chiefs of Staff exercises, and other named operational missions are all combined within theater strategy as part of persistent engagement in the steady state routine environment. Although the command relationship is the same as with larger deployments—that is, the TSOC has operational control of deployed units—the U.S. Ambassador is the designated lead U.S. Government representative for all U.S. Government activities in the HN. Although the country will coordinate with the GCC, the U.S. Ambassador has effective control over what U.S. Government activities will or will not be permitted in his or her assigned HN. Moreover, the country team is charged with understanding what the HN itself is likely to permit or deny. For all of these reasons, when SF Soldiers are operating in the country team operational environment, they concentrate on effective intergovernmental cooperation and mature personal relationship building to achieve mission success.

Chapter 6

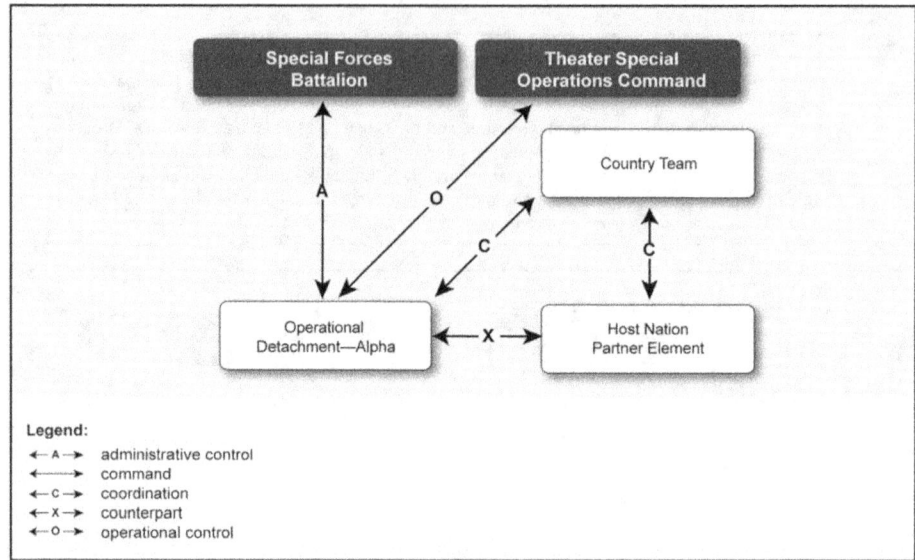

Figure 6-2. Special Forces element relationship during the conduct of theater security cooperation plan events in the steady state

Employment

6-85. Figure 6-3, page 6-17, shows a possibility for mission command and advisory assistance relationships for a single SOTF deployed to provide advisory assistance to HN forces. In Figure 6-3, the AOBs are each responsible for providing advisory assistance to a single HN brigade-sized unit, whereas the SOTF is responsible for providing advisory assistance to a HN subnational or regional mission command element.

RELATIONSHIPS WITH COUNTRY TEAM/INTERAGENCY

6-86. The country team is the point of coordination within the host country for the diplomatic mission. The purpose is to achieve a unity of effort, to coordinate, and to inform the various organizations of operations. SF units coordinate their efforts with the country team prior to entry into the HN country, during the conduct of the mission, and as an outbrief at the completion of the mission. An SF unit coordinates with individual agencies, such as the U.S. Agency for International Development or the Drug Enforcement Administration, depending on the particular issues associated with the HN.

INFORMAL RELATIONSHIP WITH NONGOVERNMENTAL ORGANIZATIONS

6-87. The SF unit may also act as the coordinating or facilitating activity for foreign humanitarian assistance provided by the international nongovernmental organizations responding to the emergency needs of a community in the FID area of operations. The SF unit should get its HN military unit counterparts involved in this activity as early as possible to foster public support for the HN military.

CONTINGENCY OPERATIONS

6-88. SF units must be able to respond rapidly to certain crises, either unilaterally or as a part of an interagency and/or multinational effort, when directed by the President or Secretary of Defense. The ability of the United States to respond rapidly with appropriate options to potential or actual crises contributes to regional stability. Thus, a joint operation may often be planned and executed as a crisis response or limited contingency. Examples of crisis response and limited contingency operations include employment of force in a terrorist attack, a single precision strike, a noncombatant evacuation operation, or a humanitarian assistance mission.

Planning Considerations for Unconventional Warfare and Foreign Internal Defense

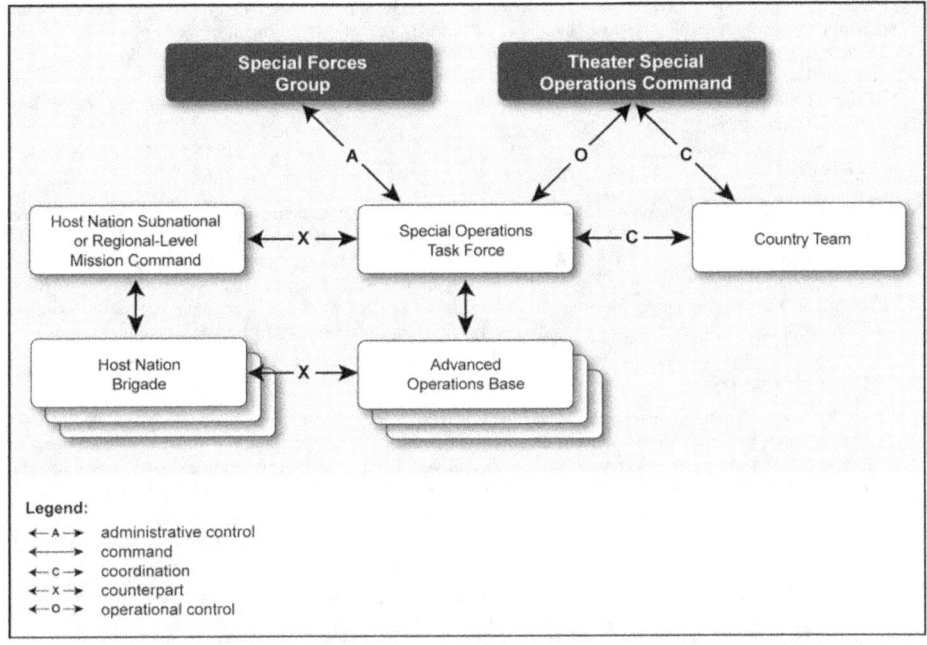

Figure 6-3. Advanced operations base and special operations task force providing advisory assistance

6-89. The crisis response force is a rapidly deployable contingency force for the GCC and, in some cases, is already forward deployed (as is the case with C Company, 3d Battalion, 1st SFG in Okinawa, Japan; and C Company, 1st Battalion, 10th SFG in Stuttgart, Germany). The operations they conduct include, but are not limited to, special reconnaissance, direct action, and counterterrorism to attain the military, political, economic, and psychological objectives of the GCC. Additionally, the crisis response force conducts planning, coordination, and execution of special operations and civil-military operations as part of a joint, combined, or unilateral force.

RAPID DEPLOYMENT

6-90. GCCs respond to developing crises at the direction of the President. If the crisis revolves around external threats to a regional partner, GCCs employ joint forces to deter aggression and signal U.S. commitment (for example, deploying joint forces to train in Kuwait). If the crisis is caused by an internal conflict that threatens regional stability, U.S. forces may intervene to restore or guarantee stability (for example, Operation RESTORE DEMOCRACY—the 1994 intervention in Haiti). If the crisis is within U.S. territory (for example, a natural or manmade disaster or deliberate attack), U.S. joint forces will conduct civil support/homeland defense operations as directed by the President and Secretary of Defense. Prompt deployment of sufficient forces in the initial phase of a crisis can preclude the need to deploy larger forces later. Effective early intervention also can deny a threat time to set conditions in their favor or achieve destabilizing objectives, or mitigate the effects of a natural or manmade disaster. Deploying a credible force rapidly is one step in deterring or blocking aggression. However, deployment alone will not guarantee success. Achieving successful deterrence involves convincing the threat that the deployed force is able to conduct decisive operations and the national leadership is willing to employ that force and to deploy more forces if necessary.

6-91. As mentioned in Chapter 5, JTF Sword is USASOC's rapid deployment JTF and possesses the capability to deploy a SOF JTF core element to facilitate the rapid establishment of a command and control element to support GCC requirements. JTF Sword is an organized, trained, and equipped deployable SOF command and control element that will primarily execute missions at the operational level for short-duration deployments (less than 180 days). It assumes command of any SOF elements that are deployed to the contingency area of responsibility.

RETASKING

6-92. SF units that are already deployed outside the continental United States can be retasked before, during, or immediately following a mission to respond to a crisis. This was the case for elements of the 7th SFG deployed throughout Central America in April 2010 when an earthquake struck Haiti. ODAs that were deployed on counterdrug operations throughout Central America were redirected to deploy to Haiti. They assisted in reconnoitering for passable roads and bridges that were still operational in order to facilitate movement of conventional force rescue and security operations throughout the remote parts of the country.

STABILITY OPERATIONS

6-93. The term "stability operations" is defined as "an overarching term encompassing various military missions, tasks, and activities conducted outside the United States in coordination with other instruments of national power to maintain or reestablish a safe and secure environment, provide essential governmental services, emergency infrastructure reconstruction, and humanitarian relief" (JP 3-0). The Army's operational concept envisions stability operations being conducted during combined arms maneuver and wide area security, and establishes or maintains a safe and secure environment and facilitates reconciliation among local or regional adversaries.

6-94. Stability operations are particularly emphasized following the achievement of major combat objectives; however, major operation and campaign plans must feature an appropriate balance between offensive, defensive, and stability operations in all phases. Even during sustained combat operations, there is a need to establish or restore civil security and control, and provide humanitarian relief as succeeding areas are occupied, bypassed, or returned to a transitional authority of HN control.

6-95. SF teams, by operating through or with the indigenous population, observe firsthand the developing threats to the HN and conventional stability efforts. This familiarity with the populace provides early warning of destabilizing activities during transitions in forces or phases of a campaign, when subversive, lawless, or insurgent actors look to seize control of the populace. By living with the population, SF Soldiers operate uncomfortably close to subversive, insurgent, or criminal elements, suppressing their

activities while stability operations achieve desired effects. By working through indigenous security forces, SF keeps a lower profile and thereby builds the popular perception as well as the reality that HN forces can secure the environment unilaterally. In wide area security (the use of combat power to protect populations, forces, infrastructure, and activities), SF, through force multiplication, closes the gaps where destabilizing elements can operate, rest, refit, and recruit. Without security, particularly HN security capacity, stability efforts will falter. SF provides commanders with an element of combat power that intrinsically builds HN security capacity and coordinates stability tasks, especially when SF teams have had a persistent presence in the region.

Chapter 7

Sustainment

The line between disorder and order lies in logistics.

Sun Tzu

The modus operandi of the guerrilla dictates a need for extreme mobility and austere and responsive logistical support. Contrary to popular belief, no guerrilla force ever operated successfully over extended periods of time without some degree of outside support—either from a sponsoring government (weapons/ammunition), the enemy (military materiel), the local population (food and shelter), or a combination of these sources. This is the way it has been in the past and there is no available evidence to indicate that such will not prevail in the future.

Marco J. Caraccia

An underestimated SOF truth is "most special operations require non-SOF support." The operational effectiveness of deployed SF cannot be achieved without being enabled by Service partners. Operational elements are the cumulative product of the Service-provided personnel and Service-common equipment enhanced by SF selection, training, and SOF modifications. Having the Service force provide the "right sized" sustainment allows SF to carry on a sustainable operation. Nevertheless, SF detachments—because of their organic capabilities (such as weapons, engineering, medicine, communications, and intelligence)—possess SOF' smallest sustainment footprint with their ability to operate in austere and remote locations. To be effective in most theaters of operation SF elements must understand organic sustainment capability as well as other enablers, to include Army and joint capabilities.

ARMY SPECIAL OPERATIONS FORCES SUPPORT TO SPECIAL FORCES OPERATIONS

7-1. On every deployment, an SF detachment should always ask who is providing their support. Each SFG possesses an organic group support battalion with a subordinate group support company and a group service support company. The group support battalion is a multifunctional logistics organization organic to the SFG with force structure and capabilities tailored to support the group. The SFG support battalion plans, coordinates, and executes sustainment operations for the group and, when directed, supports forces task organized with the group or an SF-led JSOTF. Each SF battalion has an organic support company that provides limited logistics.

7-2. The forward support company provides routine administrative and logistics support to the SF battalions. The forward support company comprises the sustainment, distribution, and maintenance platoons. The forward support company is a multifunctional logistics company providing maintenance, limited Class I through Class IX supplies, fuel and water production, ammunition holding, and transportation. The forward support company is independently deployable and capable of providing for the entire SF battalion and its attached elements. When the SF battalion establishes a SOTF, the forward support company commander may coordinate and supervise the support center logistics activities.

7-3. Because of the nature of an SF detachment's missions, sustainment (logistics) planning considerations differ somewhat from conventional Army elements. SF detachments need to be able to plan for a variety of operations, to include humanitarian, civil, and security assistance programs. Deployed SF units often operate in isolated and austere locations where distribution and resupply are key considerations that may require organic, Army, or joint capabilities, or the total dependence on indigenous forces and HN support. Sustainment planning must first consider the existing infrastructure in the joint operations area and the country or region in which they operate from or within.

7-4. When an SF unit deploys into an undeveloped theater of operations, it must bring sufficient resources to survive and operate until it establishes a support system or makes coordination for the Army Service component command, HN, or contracted support. An undeveloped theater of operations does not have a significant U.S. sustainment base. Pre-positioned war reserve materiel stock, in-theater operational project stocks, and foreign nation support agreements are minimal or nonexistent. All SF units require services to sustain food, water, and clothing, as well as medical and personnel needs. SOTFs often use a combination of external support, organic support, the 528th Sustainment Brigade, or other logistics systems to sustain their operations. SF commanders and their staffs task organize their assets to work with the logistics mechanisms existing in the theater of operations. Deployed SOF units in an undeveloped theater of operations may have to bypass normal logistics support echelons. They may maintain direct contact with their parent units in the continental United States, or they may request a tailored support package from the 528th Sustainment Brigade to accompany them into the theater of operations. The 528th Sustainment Brigade can then request directly from the continental United States wholesale logistics system and provide support and sustainment to the SOF units. Deployed SOF may also rely on Army Service component command contracting to obtain local support and sustainment. In practice, the solution may be some combination of all options.

7-5. In a developed theater of operations, the Army Service component command establishes a logistical structure within the area of responsibility that provides sustainment operations in support of ARSOF. Pre-positioned war reserve materiel stock and operational project stocks are in place, and foreign nation support agreements exist. The ARSOF logistical force structure has the mechanisms to "plug in" all joint and Army logistical and sustainment structures required for replenishment operations. It uses the same emerging technologies and support concepts as joint and Army forces.

HOST-NATION SUPPORT

7-6. Military operations often are affected by agreements between the United States and the HNs (and other nations, if the United States participates as a member of a multinational organization). These international agreements address a wide range of issues from legal jurisdiction involving crimes committed by U.S. personnel to the hiring of HN personnel to support an operation. International agreements can also influence the extent that contracting is used in support of military operations, as agreements determine a contractor's tax status, freedom of movement, immunities, and customs requirements. These are all important considerations when deciding whether to employ contractors.

7-7. The effect that international agreements might have on contracting support in a particular theater of operations must be considered during any operational planning. Because these agreements vary from nation to nation, planners must coordinate with their servicing command or theater of operations legal activity to determine if any agreements apply to the area of operations and if they would affect contracting support. Typically, international agreements that affect contracting support do so in terms of directing the use of HN support before contracting with commercial firms or restricting the commercial firms with which they can contract. In some cases, international agreements may prohibit any contracting in a specified country or region.

7-8. In planning contract support, commanders must consider the following:
- HN support, contingency contracting, and the logistics civilian augmentation program supplement (not replace) the existing logistics systems.
- The lack of any U.S. international agreements—such as HN support, inter-Service, status of forces, and other authoritative agreements in the theater of operations, or specific provisions in applicable agreements—may limit the contracting officer's ability to satisfy some requirements.
- Contract law attorneys must be deployed early to conduct legal reviews of procurements.

- U.S. public laws and the federal acquisition regulation, the defense federal acquisition regulation supplement, and the Army federal acquisition regulation supplement are not revoked or suspended by contingencies unless specifically exempted. Acquisition personnel must, therefore, comply with Federal law and applicable regulations in contingency contracting.
- Contracting, finance, and resource management remain Service responsibilities.

CONTRACTOR SUPPORT

7-9. Contingency contracting legally secures the supplies, services, and construction necessary to support missions of a deployed force. Contractors and their services in support of these missions fall into three basic categories: theater of operations support contractors, external support contractors, and systems support contractors.

THEATER OF OPERATIONS SUPPORT CONTRACTORS

7-10. Theater of operations support contractors support deployed operational forces under prearranged contracts or contracts awarded from the mission area by contracting officers serving under the direct contracting authority of the theater of operations principal assistant responsible for contracting. Theater of operations support contractors provide goods, services, and minor construction—usually from the local vendor base—to meet the immediate needs of operational commanders. Immediate contracts involve deployed contracting officers procuring goods, services, and minor construction, usually from the local vendors or nearby offshore sources, immediately before and during the operation. Theater of operations support contracting occurs according to the principal assistant responsible for contracting theater of operations contracting plan (an appendix to the operation plan or operation order), which governs all procurement of goods, services, and minor construction within the area of operations.

EXTERNAL SUPPORT CONTRACTORS

7-11. External support contractors provide support for deployed operational forces and are separate and distinct from theater of operations support or systems support contractors. Contracts may be prearranged or awarded during the contingency to support the mission.

7-12. External support contractors establish and maintain liaison with the theater of operations principal assistant responsible for contracting as they conduct their unique support missions. They procure goods and services required within the theater of operations according to the principal assistant responsible for the contracting plan (published in the operation plan or the operation order).

SYSTEMS SUPPORT CONTRACTORS

7-13. Systems support contractors support deployed operational forces under prearranged contracts awarded by program managers, program executive officers, and the United States Army Materiel Command to provide specific materiel systems throughout their lifecycle. This support is conducted during both peacetime and contingency operations. These systems include vehicles, weapon systems, aircraft, control infrastructure, and communications equipment.

CONTRACTOR SECURITY REQUIREMENTS

7-14. The nature of the contingency operation determines security requirements for the contractor's operation and personnel. Even humanitarian operations require security arrangements. As the possibility of hostilities increases, contractor security must likewise increase.

7-15. Provisions of the law of war do not consider contractor personnel and DOD civilians as combatants. To facilitate their movement and to dictate the type of treatment they should be rendered if captured, contractor and DOD civilian personnel should be issued identification cards that correctly identify them as civilians accompanying an armed force. Commanders must provide security to contractors who support their operations, or they must eliminate the use of contractor support as an option in areas where security becomes an issue.

Operational Contracting Support

7-16. SF units have no organic operational contracting capability and are reliant upon contracting support from the theater GCC, Service components, and USASOC/USSOCOM contracting activities. The key to successful contingency contracting execution in support of SF is the preplanning and early identification of the goods and services that are required. SF may have special mission requirements coupled with the diversity and complexity of assigned missions that may require an integrated approach from different contracting activities. This is especially important when working with HN vendors and indigenous sources of goods and services.

Generating Force Contracting

7-17. The USASOC Deputy Chief of Staff, Acquisition and Contracting, provides SOF contracted procurement capacity to USASOC units during force generation. The Deputy Chief of Staff, Acquisition and Contracting, contracting authority is derived from the USSOCOM. Primarily, the Deputy Chief of Staff, Acquisition and Contracting, obligates Major Force Program-11 funding through contracts written by their office in the USASOC headquarters. Installation contracting offices provide operations and maintenance contracted procurement capacity to SF units during force generation. Installation contracting office contracting authority is derived from the Mission and Installation Contracting Command. Primarily, installation contracting offices obligate Army operations and maintenance funding through contracts written at the installation they support. The Deputy Chief of Staff, Acquisition and Contracting, possesses the capacity to manage a field ordering officer program during force generation to support SF training at Fort Bragg, North Carolina.

Operational and Exercise Contracting

7-18. The lead Service for contracting supporting a GCC typically provides contracting support to SF during operations (deployments) and exercises (Joint Chiefs of Staff-sponsored or joint combined exchange training). The lead Service for contracting varies depending on the GCC and by country within some geographic combatant commands. Expeditionary Contracting Command contracting support brigades collocate with and support Army Service component commands worldwide. Depending on the theater, the Expeditionary Contracting Command contracting support brigade may be the lead Service for contracting support.

7-19. The exception to the lead Service for contracting support rule is Special Operations Component, U.S. Central Command. Special Operations Component, U.S. Central Command, is the only TSOC with expeditionary contracting capacity derived from USSOCOM. Special Operations Component, U.S. Central Command, contracting is responsible for SOF contracting and contracting support where no other contracting capacity exists in the U.S. Central Command area of responsibility. As a potential operation or exercise force provider to GCCs, the Expeditionary Contracting Command may provide contingency contracting officers to support a lead Service for contracting office or Special Operations Component, U.S. Central Command, contracting. When the Expeditionary Contracting Command is not the lead Service for contracting support in the area and a contingency contracting officer is required, support must be requested through either the Joint Capability Requirements Manager or the Joint Training Information Management System. The Joint Capability Requirements Manager is used for operational support and the Joint Training Information Management System is used for training support.

7-20. Contingency contracting support is provided to units with validated requests based on contingency contracting officer availability across the Services (among other factors). A request through either the Joint Capability Requirements Manager or the Joint Training Information Management System does not guarantee Expeditionary Contracting Command contingency contracting officer support. Lead Services for contracting and Special Operations Command, Central Command, contracting have the capability to manage field ordering officer programs during operations and exercises. Installation contracting offices have the capability to manage field ordering officer programs to support short-duration (60 days) exercises when SF units deploy from home station and return to home station. Installation contracting offices perform this function only when no other contracting support is required during the exercise.

Operational and Exercise Contracting Support Planning and Training

7-21. The Expeditionary Contracting Command functions as the operational and exercise contracting support planner and the predeployment contingency contracting support trainers to SF units. The Expeditionary Contracting Command is the USASFC advocate and liaison to bring about effective and efficient operational and exercise contracting support through coordination with geographic combatant command lead Services for contracting and Special Operations Component, U.S. Central Command, contracting. Contingency contracting unit training for contracting officer representatives and field ordering officers is available during predeployment. Expeditionary Contracting Command contingency contracting teams collocated with SFGs perform duty in installation contracting offices (including at Eglin Air Force Base, Florida). SF units requiring planning support, operational and exercise contracting support coordination, or contingency contracting unit training should contact Expeditionary Contracting Command contingency contracting teams through the installation contracting office.

STATEMENT OF REQUIREMENTS

7-22. A critical source of information the Army Service component command needs in its coordination and facilitation functions is the statement of requirements provided by the ARSOF units. The TSOC logistics staff and other logistics staffs must be proactive and included in the mission-planning process. The logistics planners must anticipate operational unit requirements at all stages of the mission. Ideally, the logistics staff uses the Army Service component command operation plan in preparing its concept plan for inclusion in the mission order. This approach allows area of responsibility support elements time to review required support before the SOF mission unit submits its mission-tailored statement of requirements. This review is especially critical in crisis-action planning and short-notice mission changes.

7-23. The statement of requirements is a living document that requires periodic reevaluation and updating as requirements change. Determination of requirements begins with the receipt of the mission. Time and accuracy are critical factors. Although contingency planning is the preferred method, crisis-action planning is within the framework. The key is to anticipate requirements based on emerging operations and then to use approved operation plans.

7-24. The intent of the statement of requirements process is to identify logistics needs early in the planning cycle. The unit or task force coordinates through its higher headquarters operations and logistics staffs to provide the USASOC Deputy Chief of Staff for Operations an initial list of requirements. The USASOC Deputy Chief of Staff for Operations tasks the Deputy Chief of Staff for Logistics to source all requirements as follows:

- When an SF unit receives a mission, it updates the standing statement of requirements developed during the deliberate-planning process. The SF commander uses this statement of requirements to cross-level supplies needed at the assigned mission-unit level. The statement of requirements identifies, consolidates, and prioritizes all unit requirements that exceed organic capabilities. The mission unit forwards it to the next-higher organization.
- At the next-higher level, the statement of requirements starts the process into the operational channels (operations staff sections). The operations and logistics sections review the statement of requirements and direct or assist cross-leveling and transfer of needed items in the most expeditious way possible. This staff level then forwards the statement of requirements to the next-higher level for any supplies and services still remaining on the statement of requirements.
- Any supplies and services still not resourced on the statement of requirements are again passed up the chain. This level forwards a statement of requirements requesting only the supplies and services not previously obtained.
- At the next level (USASOC), the requirements that can be obtained within the USASOC are coordinated and transferred. USASOC coordinates with Headquarters, Department of the Army; Army Materiel Command (subordinate commands); and other agencies and commands to source all requirements.
- To complete the statement of requirements process, USASOC forwards unsatisfied support requirements (two copies of the statement of requirements—one to the TSOC and the other to the Army Service component command for information pending validation) to the TSOC for

validation. The TSOC coordinates with the Army Service component command for the needed supplies and services.

- The Army Service component command then tasks the assigned units the sustainment mission. The Army Service component command publishes a support plan detailing how the ARSOF unit will be supported. If the Army Service component command cannot sustain or if a sister Service is better suited to sustain the ARSOF mission, the Army Service component command forwards the statement of requirements to the GCC for assistance.

7-25. The development and coordination of a unit statement of requirements is a dynamic process that concurrently occurs at multiple echelons. ARSOF develop a formal statement of requirements to support theater of operations contingency planning and contingency operations and then forward the document to the TSOC for validation. Because the statement of requirements is often a living document that requires frequent revision, and given the fluid nature of theater of operations planning, the TSOC and Army Service component command may begin coordinating new ARSOF requirements before receipt of a validated revision of the statement of requirements.

PLANNING AND EXECUTING THEATER OPERATIONS SUPPORT

7-26. The Army Service component command provides the necessary capability for the Army forces assigned to a combatant command. GCCs support SOF in their areas of responsibility. The ARSOF logistics planners identify the support requirements in the planning phase. The Army Service component command must also identify the logistics shortfalls for inclusion in the GCC's risk assessment in an area of responsibility. If the Army Service component command cannot support ARSOF, the Army Service component command must raise the shortfall to the supported GCC for resolution. The TSOC tasks missions to ARSOF. The TSOC works closely with the combatant command staff and the Army Service component command to articulate the ARSOF requirements. The GCC establishes priorities and allocates the available resources to ARSOF to accomplish each mission. The Army Service component command develops the area of responsibility support plan, which includes sustainment of ARSOF by the area of responsibility logistics organizations. The TSOC then monitors ARSOF sustainment.

ARMY SUSTAINMENT PRINCIPLES

7-27. ADRP 4-0 describes Army sustainment support to ARSOF. (ATP) 3-05.40 provides the United States ARSOF commander and staff with information on the structure and functions involved in sustainment activities. The Army retains responsibility for the sustainment of forces it allocates to a joint force. The sustainment warfighting function consists of three major sub functions: logistics, personnel services, and health services support. Logistics support covers supply, field services, maintenance, transportation, distribution, operational contract support, and general engineering support. Support providers include strategic-level providers, the generating force, and the operating forces within the Army. For example, U.S. Army Forces Command is an Army generating force that provides logistics forces to ARSOF in support of global operations through the Army force generation process.

7-28. USASOC fulfills Army Service component command-like functions as a functional Army Service component command for SF and is responsible for the manning/organization, training, equipping, administrating, and sustaining of ARSOF assigned to USSOCOM. The theater Army Service component commands exercise administrative control over Army forces within the GCC's area of responsibility and support the forces assigned, to include the TSOCs. Although Title 10, U.S. Code, support falls under each individual military Service, a combination of the different Service capabilities within an area of responsibility can create interdependent and efficient use of logistics sustainment capabilities. Sustainment of a joint force can be provided through a combination of executive agent, lead Service, and subordinate logistics command designations as described in ADRP 4-0 and JP 4-0.

7-29. SF operations are often joint in nature and support relationships are directed by the GCCs. GCCs have directive authority for logistics for their assigned forces ensuring effective execution of operation plans and economy of support forces, creating interdependent and efficient use of the different Service components' capabilities. The President or Secretary of Defense may extend directive authority for logistics to attached forces when transferring forces for GCCs' missions. Interagency coordination occurs between elements of DOD and U.S. Government agencies in order to achieve specific strategic objectives. Military operations, to include civil support operations, must be coordinated with the activities of other agencies of the U.S. Government agencies, intergovernmental organizations, nongovernmental organizations, and various HN agencies.

7-30. SF sustainment principles are an integrated process consisting of people, systems, materiel, health services, and other support. Logistics—one of the sustainment warfighting subfunctions—consists of supply, transportation, distribution, field services, maintenance, operational contract support, and engineering. Other sustainment subfunctions are personnel services and health services, financial management, legal, and religious support.

LOGISTICS

7-31. Standard basic loads are inadequate for SF operations in an undeveloped theater. For example, an SF unit may have to deploy with 30 days of supply (15-day order-to-ship time, 10-day operating level, 5-day safety level). These quantities of supplies may exceed the SFG capacity to move and store. The group and battalion logistics staffs normally divide these loads into accompanying supplies and preplanned follow-on supplies. Supply procedures for most classes of supply vary in an undeveloped theater. Except for field rations, the SFG may rely heavily upon local contract support for fresh Class I supplies and dining facility operations. To reduce demand on the logistics system, SF may purchase Class II, III, IV, and VI supplies locally or from third-party contractors. SF personnel normally receive Class V and IX supplies through the standard U.S. system, but with greater reliance on Army lines of communications.

7-32. Supply support for an ODA, once approved by the company executive officer, starts with the company supply noncommissioned officer—the commander's principal logistician. The supply sergeant is primarily responsible for supervising or performing tasks involving the general upkeep and maintenance of all Army supplies and equipment. The supply sergeant assists an ODA with coordinating and tracking support through the battalion and group headquarters.

7-33. The following list provides a breakdown of the classes of supply with accompanying symbols and the responsible section within an SF battalion:

- **Class I—Subsistence (Forward Support Company).** Subsistence is unusual because of the complexities of activities associated with its handling. Planning for Class I support is considered and synchronized from strategic to tactical levels. The result of planning and integrating Class I operations is to ensure subsistence arrives in the right quantities and proper rations are located when and where they are needed. Certain exceptions may apply under special circumstances.
- **Class II—Clothing, Individual Equipment, Tools, and Administrative Supplies (Company Supply).** The supply noncommissioned officer orders Class II through the Army supply system. Equipment that is not in the modified table of organization and equipment and nonstandard equipment/supplies are procured with assistance from the battalion logistics staff and property book office to supply the company. The Special Operations Forces Personal Equipment Advanced Requirements 2.0 program has been established in each SFG (Special Operations Forces Personal Equipment Advanced Requirements—Forward) to equip Soldiers with protective gear tailored to each individual. Each SFG is equipped with contractors to assist in this program. The individual is responsible for this equipment until leaving the SF community, at which time the individual will turn in the equipment.
- **Class III—Petroleum, Oils, and Lubricants (Forward Support Company).** Petroleum, oils, and lubricants are ordered by the forward support company.
- **Class IV—Construction Materials (Company Supply).** Class IV is ordered by the company supply section.
- **Class V—Ammunition (Operations Staff/Forward Support Company).** Requirements are validated through operational channels (battalion/group operations staff) and sourcing is through the forward support company and group logistics staff. Joint combined exchange training ammunition is used for HN forces to fire their assigned military weapon and not the weapons

brought into the HN by SF conducting the exercise. The type and amount of ammunition must be determined early on with sufficient lead time to allow for the pricing and availability through the Joint Munitions Command, the processing of the military interdepartmental purchase request, and the submission of the DA Form 581 (Request for Issue and Turn-in of Ammunition).

- **Class VI—Personal Demand Items (Company Supply/Battalion Logistics Staff).** Class VI is limited to items required for the minimum personal hygiene, comfort, and welfare of the Soldier. Items such as essential toilet articles and confections are issued in a sundry pack through Class I channels.
- **Class VII—Major End Items: Weapons, Electronics, and Vehicles (Company Supply/Battalion Property Book Office/Battalion Logistics Staff).** The supply noncommissioned officer works with the battalion logistics staff and property book office to manage company/ODA property. The SF engineer sergeant is responsible for assisting the ODA commander in managing detachment property. The ODA commander signs for a detachment's property from the company commander and then issues a sub-hand receipt to individual detachment members. Supply operations at the detachment and company level are an integral part of the group commander's ability to account for all government property entrusted to his care. Additionally, properly executed supply operations give the commander the visibility of government property to ensure the force has the necessary equipment to accomplish their assigned mission-essential task list. Deploying units develop their mission-essential equipment list upon receipt of their deployment orders. Mission-essential equipment list-required items will be designated as either theater provided (Army-common received from theater assets), ARSOF theater provided (remain in theater; these assets come from units' on-hand balances and can be Army-common or special operations-peculiar items), or to accompany troops.
- **Class VIII—Medical Materials (SF Medical Sergeant/Battalion Medical Section).** The SF medical sergeant coordinates with the battalion medical section to secure required Class VIII support.
- **Class IX—Repair Parts (Motor Pool, Emergency Medical Services).** Class IX support is coordinated by the motor pool and emergency medical services.
- **Class X—Material for Nonmilitary Programs (Battalion CA Staff).** Class IX support generally includes civil-type projects that require different funding/procurement sources.
- **Nonstandard Items.** Nonstandard items describe items that are not stocked, type-classified, or safety-approved for use by U.S. forces. The acquisition and infiltration of nonstandard items into the area of operations can be very important to help maintain low visibility of indigenous resistance forces where foreign equipment would otherwise draw unwelcomed attention.
- **Joint Operational Stocks.** The joint operational stocks are supplies of centrally managed and maintained equipment. Joint operational stocks are available on a loan basis from USSOCOM. In addition to joint operational stocks, the USASOC Redistribution Center provides the same services as the joint operational stocks with the exception that it is used only for ARSOF units.

TRANSPORTATION

7-34. SF transportation is the moving and transferring of units, personnel, equipment, and supplies to support SF missions. SF personnel have limited organic transportation resources and are very dependent upon the other Service components and strategic providers, such as the U.S. Transportation Command. Army transportation plays a key role in facilitating force projection and sustainment in most theaters of operation. The group support battalion/forward support company or support center—ground) and the 160th Special Operations Aviation Regiment (rotary wing) are for the most part the only significant organic transportation support that ARSOF have available.

7-35. Undeveloped or immature theaters have poor lines of communications in some locations; aviation assets should be coordinated to deploy early in support of ARSOF logistics operations. These aviation assets must include an adequate maintenance support package for autonomous and continuous operations. The TSOC, the 528th Sustainment Brigade, ARSOF liaison elements, and theater sustainment commands should review HN and any other possible lift assets to meet additional unresourced transportation

Sustainment

requirements. Regardless of the source of aviation assets used, this support must be dedicated (or readily available) for administrative and logistical requirements. Logistical replenishment operations conducted by the group support battalion (support center) and theater sustainment command elements provide critical for sustainment for SF teams deployed into isolated, austere, and uncertain environments.

7-36. Contingency, Joint Chiefs of Staff, and joint combined exchange training events are planned and resourced using the Joint Operation Planning and Execution System. The Joint Operation Planning and Execution System produces a product known as the time-phased force and deployment data, which is used by U.S. Transportation Command to source transportation assets to support the overall mission. The method of populating Joint Operation Planning and Execution System with the required information varies based on the operation. Joint Operation Planning and Execution System information is provided through operational channels. The USASOC Deputy Chief of Staff for Operations is the lead agency for Joint Operation Planning and Execution System management and continually updates policies and procedures to support this function.

7-37. Operations, exercises, and training events that do not require the use of Joint Operation Planning and Execution System may be supported using other DOD systems that manage smaller operations. For the most part, these internal decentralized missions are planned, coordinated, and executed using the unit's operation and maintenance funds. The supporting installation transportation office can provide further guidance and assistance in executing these smaller operations. A brief description of the different types of capabilities are listed below:

- **Strategic Air.** Strategic air is a common-user airlift linking theaters to the continental United States and to other theaters, as well as airlift within the continental United States. These airlift assets are assigned to the Commander, U.S. Transportation Command. Because of the inter-theater range usually involved, strategic airlift comprises heavy, longer-range, intercontinental airlift assets, but also may be augmented with shorter-range aircraft.
- **Channel Airlift.** Channel airlift is a common-user airlift providing regularly scheduled airlift for movement of sustainment cargo, depending on volume of workload, between designated aerial ports of embarkation and aerial ports of debarkation over validated contingency or distribution channel routes. In some circumstances, channel air may be used for the deployment and redeployment of SF personnel and equipment other than sustainment cargo.
- **Special Assignment Airlift Missions.** Special assignment airlift missions are airlifts supporting requirements for special pickup or delivery by the U.S. Air Force Air Mobility Command at points other than established Air Mobility Command routes, and which require special consideration because of the number of passengers involved, the weight or size of the cargo, the urgency or sensitivity of movement, or other special factors.
- **Charter/Contracted Airlift.** Chartered/contracted airlifts are missions that fall under the umbrella of strategic, channel, or special assignment airlift mission policies and procedures. They are supported by U.S. flag carriers and are contracted by U.S. Transportation Command to perform specific missions. The majority of these missions involve the movement of personnel from continental United States to outside the continental United States locations; however, cargo-carrying assets may also be used when DOD assets are limited.
- **Nonstandard Aviation.** Nonstandard aviation is a specific mission capability that provides SF freedom of movement and access to austere operating locations while maintaining a civilian appearance to casual observers and the local population. Nonstandard aviation provides capability in areas where an obvious U.S. military presence and footprint may be denied. In some cases, intratheater air mobility is coordinated through a dedicated SOF command and control structure assigned to Air Force Special Operations Command. Nonstandard aviation provides theater-focused, low-visibility, short takeoff and landing, and other aviation assets to support the mission roadmap guidance (for example, overseas contingency operations and irregular warfare).

7-38. Although the majority of movements within the ARSOF community are executed by air, SF must be able to execute surface movement (water, rail, and road). In order to accomplish this end state, the SFG must rely on the support provided by their installation transportation office. Each installation transportation office has functional experts in unit movements; container management; port, convoy, and rail operations; special assignment airlift missions; and commercial movements. Each of these areas has its own support system that should be used in order to acquire the necessary support. Each SF unit must appoint a unit movement officer to assist in the utilization of the systems and the execution of the movement requirement.

DISTRIBUTION

7-39. Distribution is the operational process of synchronizing all elements of the logistics system to deliver the right things to the right place at the right time to support the GCC. It incorporates not only transportation but also provides distribution management and asset visibility activities. The group support battalion's movement control team can provide movement guidance and assistance for both internal (organic) and external (conventional forces) assets allowing for the movement of personnel and supplies by air, ground, or sea. This team also plans, coordinates, and executes the requirements for moving units to continental United States and outside the continental United States locations.

7-40. ODA personnel request supplies based on operational needs through a logistics situation report once communications are established with the SOTF/CJSOTF. These supplies consist of major equipment items that units do not consume at a predictable rate. The theater sustainment capabilities (theater sustainment command, expeditionary sustainment command, and sustainment brigade), the group support battalions, and the SOTF/JSOTF logistics nodes may maintain on-hand stocks for immediate delivery as requested. It is essential that SF personnel control and maintain access to sensitive items, such as weapons, ammunition, demolitions, radios, drugs, or special equipment. An ODA member must be present at all deliveries of external supplies to maintain positive control and accountability.

Resupply Methods

7-41. The SOTF or AOB staff establishes supply levels for each class of supply in the joint special operations area. It then determines the sequence, method, and timing of delivery. The SOTF or AOB plans for three types of resupply missions: automatic, emergency, and on-call. The SOTF logistics staff requisitions the supplies and equipment for these missions to the group support battalion. The group support battalion passes any requirement outside their capability to the nearest conventional forces sustainment element. Resources are then provided by the theater sustainment command and coordinated by the group support battalion and receiving unit. ODAs normally preplan for resupply missions while conducting mission planning.

Automatic Resupply

7-42. Automatic resupply provides items that cannot accompany the ODA during infiltration. Automatic resupply provides essential subsistence, training, and operational supplies to the ODA and its indigenous force on a preset schedule. The delivery time, location, contents, identification marking system, and authentication are preplanned. The SOTF sends supplies automatically unless the deployed ODA cancels, modifies, or reschedules the delivery.

Emergency Resupply

7-43. Emergency resupply is limited to mission-essential equipment and supplies to restore the operational capability or survivability of the ODA and its indigenous force. The group support battalion support operations staff at the JSOTF/CJSOTF executes an emergency resupply when notified by the operations staff section.

On-Call Resupply

7-44. On-call resupply provides equipment and supplies to the deployed ODA to meet operational requirements that cannot be carried during infiltration or to replace equipment lost or damaged during the operation. The deploying element, rigger section, and logistics staff section prepack and rig on-call resupply bundles. The bundles are then held in a secure location and delivered when requested by the ODA.

Sustainment

Caches

7-45. Caches are another alternative form of resupply. ODAs can stockpile materiel within the joint special operations area to support future operations or contingencies. They can also recover caches emplaced by other units from previous operations. Using caches from previous operations must be coordinated with the commander controlling the joint special operations area.

Aerial Delivery

7-46. Aerial delivery is a vital means of sustaining SF operational elements. SF elements are extremely limited in their ability to provide organic ground distribution capability; however, their ability to provide aerial delivery as a primary means of logistical support is, for the most part, unlimited.

7-47. The primary means of delivering sustainment in denied areas is aerial delivery. As a vital component in sustainment of SF, it is no longer the last resort, but rather a viable and required mode of distribution within the operational environment. Aerial delivery enables noncontiguous operations, thereby reducing the logistics footprint for SF operational elements, exposure, and risk to sustainment assets.

7-48. The group support battalion is the primary means of support for aerial delivery to SF operational elements in undeveloped and immature theaters of operations. The group support battalion facilitates rapid, no-notice deployments with initial (accompany), follow-up, sustainment (demand supported), and emergency aerial delivery support, including cargo and personnel parachute requirements (static-line and military free-fall). The major limitation to SF aerial delivery in an operational environment is aircraft availability.

7-49. There is a growing array of aerial delivery systems and methods. The basics are low-velocity (slow descent rate), high-velocity (high descent rate), low-altitude, high-altitude, and high-altitude offset joint precision airdrop system delivery. Low cost variants of low- and high-velocity parachutes are also used with the advantage of no requirement of retrograde for aerial delivery equipment. Normally, the air delivery section within the group support battalion and the supporting aircraft element determine which system and method is used after a threat analysis is conducted for the intended drop zone or if the receiving operational element has specific clandestine requirements. These needs drive the selection of aircraft, methods, and rigging equipment used.

Field Services

7-50. The SFG and SF battalion have limited organic assets to provide field services. Additional field services support can be requested from the 528th Sustainment Brigade or the conventional force theater sustainment command. Field services include mortuary affairs, airdrop, clothing exchange and bath, laundry, bread baking, textile and clothing renovation, and salvage. Mortuary affairs and airdrop are primary field services and are essential to the sustainment of combat operations. All others are secondary field services. The group support battalion has no organic mortuary affairs assets and requires augmentation or designates one noncommissioned officer within the group support battalion support operations staff section or medical platoon to coordinate mortuary affairs operations.

7-51. The USASOC bare-base sets are air-transportable modular systems. Each system is configured into different modules (tent module, mission command module, housing module, and support module) and is a complete turnkey operation, to include power generation, heating and air, and lighting capability. The tent and mission command module configuration provides workspace and equipment (tents, tables, chairs, file cabinets, briefing tent, and so on) for SOTF- and JSOTF-sized operations. These modular systems afford unit commanders a rapidly deployable package of equipment designed to provide limited life support and work areas in a bare-base environment. Modules do not include kitchen equipment. Units must provide their own manpower to set up and maintain the modules.

Maintenance

7-52. Maintenance is a challenge regardless of whether it is a developed or undeveloped theater. Many U.S. systems are complex and require special training to repair. Preventive maintenance checks and services are important and must be performed at the operator level regardless of the equipment type. If repair parts are needed, they are delivered based on logistics situation report requirements, size, and delivery methods available. Equipment needing other than field-level maintenance may be evacuated to the rear, if possible, and replaced on a one-for-one basis during logistics resupply missions. Relationships built between SF and conventional forces who are collocated can expedite equipment repairs serviced by conventional force maintenance teams so that equipment does not have to be evacuated to the rear. When feasible, HN contracts should be utilized. Most special operations-peculiar items cannot be repaired by organic Army maintenance personnel. The SFG must rely on USSOCOM civilian technicians and SF personnel who have attended special operations-peculiar maintenance training. Such equipment may require evacuation to the continental United States for repair at the manufacturer or other selected facility.

7-53. The unit forward support company's mechanical maintenance section performs consolidated field-level maintenance of wheeled vehicles and power-generation equipment along with vehicle recovery operations. The electronics maintenance section performs field-level maintenance of all signal equipment. It also performs limited sustainment-level maintenance on special operations-peculiar signal equipment, supported by USSOCOM's special operations-peculiar programs. Unit armorers perform field-level maintenance of small arms, to include special operations-peculiar weapon systems. USASOC assists deployed units with field training teams set up in combat zones to repair and replace special operations-peculiar equipment, as needed.

7-54. Required maintenance on an item of equipment may exceed unit capabilities. In this case, the mechanical maintenance or electronics maintenance section evacuates the equipment to the supporting field-level maintenance platoon (or USSOCOM special operations-peculiar maintainers). Field-level maintenance requirements are evacuated to intermediate sustainment-level maintenance units for those items that it cannot repair. Field-level maintenance also provides direct exchange service and maintains a limited number of operational readiness floats.

7-55. The two exceptions to the above stated procedures are the rigger air delivery section and the medical section. The rigger air delivery section evacuates unserviceable airdrop equipment to the theater sustainment command airdrop equipment repair and supply company. The medical section evacuates unserviceable medical equipment to the supporting medical treatment facility or medical logistics unit.

7-56. Preventive maintenance checks and services are critical regardless of operational environmental condition or region. The frequency of periodic services often differs from region to region. Repair facilities in an undeveloped theater are often unavailable. The SFG commander reviews the modified table of organization and equipment to determine the items needed to meet increased maintenance demands caused by dispersion of resources for operations. The SFG should identify maintenance support in the statement of requirements before deployment.

7-57. Prior to deploying to combat, commanders must be proactive in programming their equipment resets upon their return because of the competing resources and units. The unit's support operations staff section should coordinate with their local lifecycle management commands to schedule and lock in their projected dates. Failure to schedule the unit's reset in advance negatively affects the unit during their redeployment, refit, and recover cycle. Reset teams consist of the following:

- Small-arms repair evaluation team.
- Chemical and biological equipment repair team.
- Communications security/electronic encryption equipment repair team.
- Communication-electronics evaluation repair team.
- SOF weapons repair team.
- SOF communications repair team.
- Unmanned aircraft system repair team.

Engineering Support

7-58. SF units have engineer support requirements similar to those of conventional units; however, the nature of SF missions presents a unique set of challenges for engineer units to provide support. These challenges include remote sites, hostile situations, limited resources and equipment, and limited transportation.

7-59. Modern modular engineer units are fully capable of meeting the challenge of support to SF units, and can do so especially well when working in concert with other logistics and sustainment assets. The specific mission requirements include protection, vertical and horizontal construction, and special engineer capabilities of survey and design, construction management, well drilling, firefighting, and power generation. Modular engineer units provide the full range of capabilities to support these missions. Additionally, an engineer battalion headquarters can execute mission command and specialized support to the deployed engineer companies. When conventional forces are not in the position to provide area support, it is critical to request a route clearance package.

7-60. In terms of providing forces to perform engineer missions, each modular engineer unit performs a specific set of tasks. For example, a vertical engineer company performs construction of buildings as well as electrical, plumbing, and heating, ventilation, and air conditioning tasks. Horizontal construction (roads and airfields) is performed by horizontal companies and engineer support companies. The aspect of modularity provides the flexibility in the deployment and employment of engineer units.

7-61. One possible scenario for engineering support consists of a JSOTF operating in a theater with several battalion base camps, multiple company base camps, and numerous ODA base camps in a high-threat environment. This scenario includes a full complement of enablers—aviation, military intelligence, and logistics elements. The support requirements range from light building construction to fire base construction. Also in this scenario, conventional forces are not providing area support for route clearance based on time-distance factors. The recommended engineer support package would be an engineer battalion headquarters (minus) (Deployable Command Post 1) with a supply and maintenance element, one vertical company, one engineer support company, one well-drilling team, and a route clearance platoon. This package provides a mission command element, a small specialized support element, seven engineer platoons, and one specialized team to perform missions. The route clearance platoon allows the construction elements to travel to their project sites.

> *Note*: A route clearance platoon conducts route clearance with an attached explosive ordnance disposal unit since a route clearance platoon cannot clear certain types and sizes of munitions during their mission. The explosive ordnance disposal unit has approval to clear homemade explosives up to a certain limit if they are certified through the U.S. Army Engineer School. An explosive ordnance disposal unit is required to support the route clearance platoons.

7-62. Another possible scenario consists of a SOTF in a theater with several company base camps and numerous ODA base camps in a low to moderate threat environment. The engineer support package should consist of a vertical company (minus) with a supply and maintenance element task organized with two vertical platoons and a rapidly deployable earthmoving platoon–light from the expeditionary sustainment command. This package provides sufficient engineer assets to perform vertical and horizontal construction in support of a reinforced battalion-level organization.

PERSONNEL SERVICES

7-63. Personnel services for a deployment are broken into four phases: predeployment requirements, deployment personnel actions, redeployment, and postdeployment. Predeployment requirements vary based on the deployment notification timeline. At a minimum, a Soldier's DD Form 93 (Record of Emergency Data) and SGLV 8286 (Servicemembers' Group Life Insurance Election and Certificate) are reviewed and updated. In addition, deployment timelines must be provided to the SF unit's personnel staff section for accurate accountability. Additional requirements are provided by the SF unit's personnel staff section based on the varying degree of each deployment.

7-64. Deployed personnel actions are processed through the SF unit's personnel staff section unless otherwise stated. It is mandatory that Soldier accountability be enforced. Once redeployment preparations have started, Soldiers need to keep in mind time-sensitive actions, such as special pays and entitlements, as well as awards or evaluations that may need to be completed.

7-65. Once Soldiers have redeployed, they must complete all reintegration personnel requirements within the SF unit personnel staff section's established timelines. Specific actions include reviewing the DD Form 93 and SGLV 8286, personnel tempo, dwell time updates, and personnel record updates. These areas are emphasized since these actions are the foundation for personnel and deployment management.

FORCE HEALTH PROTECTION

7-66. A force health protection/health service support planner from the SFG and/or SF battalion medical sections must be involved as early as possible in the planning process. The force health protection/health service support planner must produce a straightforward plan without complication to ensure a continuum of care to the full range of SF operational environments. FM 4-02.43 provides additional information.

7-67. The SF medical planner must plan for all health service support functional support areas regardless of the mission or the operational environment because SF units and personnel operate over a wide area and in isolated and austere locations with limited health service support. Force health protection planning for SF assets involves numerous considerations that do not apply to conventional forces.

7-68. SF elements have Echelon I (Role 1—first level of medical care an SF Soldier would receive) capability and depend on area and theater force health protection assets for health service support requirements. SF medical assets, in coordination with conventional force health protection planners, determine what support is organic and what area/theater medical assets are provided. Early coordination and communication is the key to success for force health protection support to SF operations. Table 7-1, page 7-15, depicts the organic medical personnel structure within an SFG.

Special Forces Medical Sergeant

7-69. The SF medical sergeant forms the backbone of medical care within the SFG. There are two SF medical sergeants in each ODA. SF medical sergeants—
- Provide, train, and advise detachment members, multinational and coalition forces, or indigenous personnel in emergency and routine medical care, emergency dental care, and veterinary care.
- Establish field medical treatment facilities to support detachment operations and prepare the medical portion of area studies, operation plans, and operation orders.
- Conduct medical intelligence analysis and prepare health threat and counter-threat briefings and lessons-learned briefings.
- Requisition, assemble, and maintain detachment medical equipment and supplies.
- Supervise routine and emergency medical activities in a field or in a UW environment.

7-70. USASOC Regulation 350-1, *Army Special Operations Forces Active Component and Reserve Component Training*, addresses medical sustainment training requirements. Clinical capabilities for the SF medical sergeant are discussed in AR 40-68.

Medical Considerations

7-71. The following characteristics for SF operations must be factored into the plan:
- **Small Units and Austere Force Health Protection Capability.** SF unit locations (geographical factors, time-distance factors) may require collocation of assets and support on an area basis.
- **Operations in a Joint, Multinational, and Coalition Environment.** Operations in these environments require SF medical personnel to have a thorough knowledge of other Service component, multinational, and/or coalition forces' health service support capabilities, limitations, organization, procedures, and national caveats.

Sustainment

- **Remote Operating Areas and Long Evacuation Routes.** SF elements often operate in areas that impede evacuation by rotary-wing aircraft or where aviation assets are not available. This places a premium on the early application of trauma management and casualty stabilization.
- **Medical Evacuation, Medical Regulating, and Casualty Tracking.** Medical regulating and casualty tracking requires an understanding of SF missions and the limited availability of replacements. Leaders must account for sensitive equipment and documents if the casualty still possesses them when evacuated.

Table 7-1. Medical personnel structure within the Special Forces group

Title	Military Occupational Specialty	Grade
Special Forces Group Medical Section (Headquarters)		
Group Surgeon	61N	O-5
Psychologists	73B	O-4
Medical Operations Officer	70H	O-3
Medical Operations Sergeant	18Z	E-8
Medical Logistics Sergeant	68W	E-6
Special Forces Medical Platoon (Group Support Battalion)		
Physician Assistant	65D	O-3
Medical Logistics Officer	70K	O-3
Environmental Science Officer	72D	O-3
Special Forces Medical Sergeant	18D	E-7/E-6
Special Operations Combat Medic	68WW1	E-5
Veterinary Officer	64A	O-3
Dental Officer	63A	O-3
Dental Specialist	68E	E-4
Preventive Medicine Sergeant	68S	E-7
Physical Therapist	65B	O-3
Battalion Medical Section		
Battalion Surgeon	61N	O-4
Physician Assistant	65D	O-3
Preventive Medicine Sergeant	68S	E-7
Special Forces Medical Sergeant	18D	E-7
Special Forces Operational Detachment—Bravo		
Special Forces Medical Sergeant	18D	E-7
Special Forces Operational Detachment—Alpha		
Special Forces Medical Sergeant	18D	E-7/E-6

7-72. It is imperative that SF medical sergeants conduct operational medical planning and geographical and environmental medical threat analysis prior to all deployments to determine materiel quantities and specific additional support requirements for that mission. Medics must fully understand the SF principal tasks and the operational, tactical, and geographical constraints associated with those tasks. In planning and coordination for medical support (both internal and external), SF medical personnel must analyze operational mission planning factors against the ten medical functional areas to determine materiel and equipment requirements and develop the mission-specific medical support plan. The 10 medical functional areas are—

- Casualty evacuation.
- Medical treatment.
- Hospitalization.
- Medical logistics.

- Preventive medicine.
- Veterinary services.
- Dental services.
- Behavioral health.
- Laboratory service.
- Medical mission command systems.

7-73. Once deployed, the SF medical personnel must continually refine the medical support plan by conducting theater medical support analysis and modify the plan as theater medical support assets become available. Additional considerations when working with indigenous populations include local medical infrastructure, theater medical rules of engagement (combatant or noncombatant personnel), cultural beliefs, and effects on the population.

FINANCIAL MANAGEMENT SUPPORT

7-74. The financial management mission is to ensure that proper financial resources are available to accomplish the mission according to commanders' priorities. The primary purpose of financial management is to sustain and support operations until successful mission accomplishment. Financial management is composed of two distinct but mutually supporting functions: resource management and finance operations. Although independent of one another, these two functions must be integrated into mission planning and execution at every level. Integration facilitates the optimal allocation of financial resources to accomplish the mission. The combined efforts of resource management and finance operations work to extend Army forces' operational reach and prolong operational endurance, thereby allowing commanders to accept risk and create opportunities for decisive results.

Resource Management Support

7-75. The resource management section is the commander's leading representative responsible for financial management support. Resource management personnel advise the appropriate allocation and use of scarce resources, to include funding, in the accomplishment of the commander's assigned missions. Resource management personnel assist commanders by providing a critical capability, which matches legal and appropriate sources of funds with thoroughly vetted and valid requirements. Funding support provides flexibility through nonlethal methods to augment and, in some cases, lead the effort in obtaining the effects the commander is trying to achieve.

7-76. The resource management mission is to analyze resource requirements ensuring commanders are aware of existing resource implications in order for them to make resource informed decisions, and then obtain the necessary funding that allows the SF commander to accomplish the overall unit mission. Key resource management tasks include the following:
- Providing advice and recommendations to the commander.
- Identifying sources of funds.
- Forecasting, capturing, analyzing, and managing costs.
- Acquiring funds.
- Distributing and controlling funds.
- Tracking costs and obligations.
- Establishing and managing reimbursement processes.
- Establishing and managing the Army Managers' Internal Control Program.

7-77. Resource management personnel also provide a variety of organic support to commanders for overseas contingency operations, Joint Chiefs of Staff exercises, joint combined exchange training, and counternarcotics training.

Sustainment

Operational Funds Support

7-78. Resource management organic support includes use of operational funds. USASOC operational funds are governed by USASOC Policy Number 32-09, *Operational Funds*. Operational funds can be requested before, during, or after deployments and authorized training events. Units must work with their resource management office to request an operational fund. In general, the commander appoints a pay agent on an additional duty appointment order. This appointment authorizes the pay agent to disburse public currency according to the special instructions stated in the appointment and the written instructions provided by the financial management commander. The field ordering officer, whom the pay agent supports, receives separate instructions from contracting officials. Field ordering officers and pay agents train and work as a team; the pay agent should participate with their field ordering officer in training and vice versa. The pay agent or field ordering officer may be held personally liable for any payment not in accordance with the appointment orders or prescribed instructions. The pay agent cannot simultaneously serve as either a certifying officer or field ordering officer. The pay agent uses an official credit or debit card to make payments whenever possible. When it is not possible to use an official credit or debit card to make payments, the pay agent takes the following actions:

- Reviews all Standard Forms 44 (U.S. Government Purchase Order-Invoice-Voucher) prepared by the ordering officer.
- Disburses currency for the goods or services as stated on the Standard Form 44, but only after this form has been approved by a field ordering officer.
- Pays for purchases not to exceed established limits. An agent may not split purchases between two or more vouchers to circumvent the established limit.
- Clears the account with the disbursing officer that advanced the funds.

Other Funding Support

7-79. Funding support is a complex endeavor and requires resource management personnel to leverage multiple appropriations. Some of these appropriations are initially provided for peacetime support, along with appropriations that are newly created by Congress specifically for an operation. Commanders and resource management personnel need a thorough understanding of the statutes and regulations that govern the use of appropriated and nonappropriated funding. Resource management personnel must work closely with the fiscal lawyer to ensure compliance with fiscal requirements established by law. The following paragraphs highlight the basic appropriations that fund SOF. Multitudes of funding options are available and may include funding sources from other U.S. agencies (for example, intelligence funding, counterdrug funding, and Department of State funding). Funding authorities the financial management personnel may leverage before, during, and after contingency operations include the following:

- **Operations and Maintenance, Army (Major Force Program—2).** SF units receive some direct funding (Major Force Program—2) from the Army for some Army-common requirements. SF units use Major Force Program—2 to pay for the day-to-day expenses in garrison and during exercises, deployments, and military operations. There are threshold dollar limitations for certain types of expenditures, such as purchases of major end items of equipment and construction of permanent facilities. Operations and Maintenance, Army, is typically a one-year appropriation and must be obligated in that fiscal year (from 1 October to 30 September).
- **Operations and Maintenance, Defense (Major Force Program—11).** SF units use Major Force Program-11 for training, equipping, and employing SF with special operations-peculiar equipment, materials, supplies, and services. Major Force Program—11 is for SOF requirements only. The same dollar limitations apply to Major Force Program—11 as Major Force Program-2 funds.
- **Military Personnel, Army.** Military Personnel, Army, funding is used for pay, allowances, individual clothing, subsistence, interest on deposits, gratuities, and permanent change of station travel (including all expenses for organizational movements) for members of the Regular Army and mobilized Reserve and National Guard Soldiers. Military Personnel, Army, funding is generally available for one fiscal year and is centrally managed and funded. Since Military

Personnel, Army, funding is centrally managed, personnel should plan for the use of Military Personnel, Army, funding to ensure receipt in time to satisfy the requirement.
- **Procurement.** Whereas Operations and Maintenance, Army, funds day-to-day operations, procurement is typically used for centrally managed items or systems that are considered investment items. These items require the use of procurement funds regardless of cost (or the cost of individual components). Such items can include large pieces of equipment or systems that exceed the expense investment threshold.
- **Research, Development, Test, and Evaluation.** Research, development, test, and evaluation funds provide for the development, engineering, design, purchase, fabrication, or modification of end items, weapons, equipment, or materials. This is not an appropriation normally used in the theater by deployed units unless involved in the research, development, acquisition, and testing process. Research, development, test, and evaluation funding is available for two years.
- **Military Construction.** Military construction provides for the acquisition of land and construction of buildings for which authorizing legislation is required.

7-80. In addition to the funding support provided by the sources listed above, both commander and resource management personnel may leverage the additional funding sources listed in the following paragraphs in support of the mission.

Commanders' Emergency Response Program

7-81. The purpose of the Commanders' Emergency Response Program is to enable military commanders to respond to urgent humanitarian relief and reconstruction requirements within their areas of operation by carrying out programs that immediately assist the local populace. The program is designed to allow commanders down to the SF battalion level to have the ability to make an immediate, positive impact in their areas of operation. This authority is authorized through the enactment of annual authorization/appropriations acts and is not codified in law. DOD Financial Management Regulation 7000.14-R, Volume 12, *Special Accounts Funds and Programs*, Chapter 27 (Commanders' Emergency Response Program), provides implementing policy and guidance for the use of Commanders' Emergency Response Program funds. The guidance primarily assigns administration responsibilities, defines proper Commanders' Emergency Response Program projects, and specifies accountability procedures. This guidance is mandatory reading for anyone intending to use Commanders' Emergency Response Program funds.

Department of Defense Rewards Program

7-82. The DOD Rewards Program is not an intelligence program and is not intended to replace existing programs. DOD Financial Management Regulation 7000.14-R, Volume 12, *Special Accounts Funds and Programs*, Chapter 17 (DOD Rewards Program), provides overall policy and guidance for the implementation of the DOD Rewards Program.

7-83. Title 10, U.S. Code, Section 127b, *Assistance in Combating Terrorism: Rewards*, authorizes the DOD to pay rewards to persons for providing U.S. Government personnel or government personnel of multinational forces participating in a multinational operation with U.S. armed forces with information or nonlethal assistance that is beneficial to—
- An operation or activity of the armed forces or of multinational forces participating in a multinational operation with multinational forces conducted outside of the United States against international terrorism.
- Personnel protection of the armed forces or multinational forces participating in a combined operation with U.S. Armed Forces.

7-84. This authority is useful to encourage the local citizens of foreign countries to provide information and other assistance, including the delivery of dangerous personnel and weapons, to U.S. Government personnel or government personnel of multinational forces. GCCs provide additional policy guidance for this program within their respective area of responsibility. U.S. or multinational units pay rewards for information helpful to the multinational forces and are not limited only to information leading to the capture of a high-value individual or seizure of weapons.

Sustainment

Emergency and Extraordinary Expense Authority

7-85. Emergency and extraordinary expense authority is found in Title 10, U.S. Code, Section 127, *Emergency and Extraordinary Expenses*. This provides the Secretary of Defense and Service secretaries authority to expend operations and maintenance funds without regard to contracting and purpose limitations. This authority is provided annually in the operations and maintenance appropriations. Emergency and extraordinary expense funds are those that may be used to support certain requirements of operations. USSOCOM regulations cover these funds and define the types of acceptable expenditures. Very small amounts of this authority exist. The GCC can request the Service component to provide emergency and extraordinary expense funds. This authority does not provide cash or foreign currency to conduct an activity; rather, it provides the capability to obligate operations and maintenance funds for an activity normally not authorized for operations and maintenance funding. The USSOCOM receives emergency and extraordinary expense authority to support Official Representation Funds and Confidential Military Purpose Funds. Official Representation Funds are funds used by high-level commanders (usually division commander and above) to uphold the standing and prestige of the United States by extending official courtesies to certain officials and dignitaries of the United States and foreign countries. Used correctly, Official Representation Funds are very helpful in building relationships in contingency operations. DOD Instruction 7250.13, *Use of Appropriated Funds for Official Representation Purposes*, provides information regarding proper obligation and expenditure of these funds. Confidential Military Purpose Funds are used for operational preparation of the environment, to include advanced special operations.

Combatant Commander Initiative Fund

7-86. The Combatant Commander Initiative Fund provides a means for GCCs to react to unexpected contingencies and opportunities. Funds may be used for—

- Command and control.
- Joint exercises.
- Humanitarian and civic assistance.
- Military education and training to military and related civilian personnel of foreign countries.
- Personnel expenses of defense personnel participating in bilateral or regional cooperation programs and contingencies.
- Selected operations.

National Defense Authorization Act Authority

7-87. The National Defense Authorization Act authority is used to conduct or support programs globally that build the capacity of a foreign country's military and maritime security forces. Types of equipment provided by this funding include—

- Radios and telecommunication systems.
- Surveillance and reconnaissance systems.
- Trucks, ambulances, boats, and other vehicles.
- Small arms and rifles.
- Night vision goggles and sights.
- Clothing.

7-88. The National Defense Authorization Act authorizes assistance to foreign forces, irregular forces, groups, or individuals supporting U.S. counterterrorism military operations. This legislation also authorizes expenditure of funds to support—

- Operational preparation of the environment.
- Advanced special operations.
- Advanced force operations.
- Direct action in support of ongoing operations.

Lastly, the National Defense Authorization Act authorizes the DOD to reimburse foreign forces, groups, or individuals supporting or facilitating ongoing counterterrorism military operations by SOF.

Sensitive Mission Funds

7-89. Sensitive Mission Funds are used when the mission requires the ability to conceal the payee, the payer, or the purpose of the payment using special financial and accounting procedures.

Memorandum of Agreement

7-90. Memorandums of agreement are agreements between countries or eligible organizations that delineate responsibilities among the participants. Among these responsibilities are the participants' financial liabilities for support. These agreements define the specific mechanisms required for reimbursement of costs. Memorandums of agreement must be signed by the USASOC Chief of Staff and be based on specific legal authority and negotiated according to proper procedures.

LEGAL SUPPORT SERVICES

7-91. Command judge advocates and their paralegals are assigned at the SFG and battalion and are special staff for the commander. They provide legal advice on the six core legal disciplines—

- **Military Justice.** For military justice, the command judge advocate advises commanders with Uniform Code of Military Justice authority (company commander and above) on the administration of military justice and adverse administrative actions.
- **International and Operational Law.** International law is the application of international agreements, U.S. and foreign law, and customs related to military operations and activities. Operational law encompasses the law of war but goes beyond the traditional international law concerns to incorporate all relevant aspects of military law that affect the conduct of operations.
- **Administrative and Civil Law.** Administrative and civil law is that body of law containing the statutes, regulations, and judicial decisions that govern the establishment, functioning, and command of military organizations as well as the duties of military organizations and installations with regard to civil authorities.
- **Contract and Fiscal Law.** Contract law is the application of domestic and international law to the acquisition of goods, services, and construction. Fiscal law is the application of domestic statutes and regulations to the funding of military operations and support to nonfederal agencies and organizations.
- **Claims.** The Army claims program investigates processes, adjudicates, and settles certain claims on behalf of and against the United States (for example, property damage or loss and personal injury).
- **Legal Assistance.** Legal assistance is the provision of personal civil legal services to Soldiers, their dependents, and other eligible personnel.

Source Notes

This section lists sources by page number. Where material appears in a paragraph, it lists both page number and paragraph number.

1-1	"In any campaign...": Text is extracted from Special Text 31-90-1, *Operations Against Guerrilla Forces,* (Fort Benning, Georgia: The Infantry School, September 1950) [obsolete]; however, this document was written entirely by Colonel Russell W. Volckmann.
1-4	1-19. "envisioned Special Forces...": Alfred H. Paddock, Jr., *Psychological and Unconventional Warfare, 1941–1952: Origins of a "Special Warfare" Capability for the United States Army* (U.S. Army War College, 1979).
1-7	1-31. "the only precursor...": Thomas K. Adams, *U.S. Special Operations in Action: The Challenge of Unconventional Warfare* (New York: Routledge, 1998) 62.
1-7	"Whatever your position...": President John F. Kennedy, Address at U.S. Military Academy, West Point, New York, 6 June 1962. [Online] Available: http://www.jfklibrary.org/Asset-Viewer/Archives/JFKPOF-038-035.aspx
1-9	"I know that...": President John F. Kennedy, Message to General Yarborough, 11 April 1962. [Online] Available: http://www.jfklibrary.org/JFK/JFK-Legacy/Green-Berets.aspx
2-1	"This is another...": President John F. Kennedy, Address at U.S. Military Academy, West Point, New York, 6 June 1962. [Online] Available: http://www.jfklibrary.org/Asset-Viewer/Archives/JFKPOF-038-035.aspx
3-1	"Unconventional warfare... remains...": Robert M. Gates, Remarks at the Dedication of the OSS Memorial, Langley, VA, 12 June 1992. [Online] Available: http://smallwarsjournal.com/blog/journal/docs-temp/642-cochran.pdf
4-1	"Do not try...": T.E. Lawrence, "Twenty-Seven Articles," *Arab Bulletin* (20 August 1917): 126. [Online] Available: http://wwi.lib.byu.edu/index.php/The_27_Articles_of_T.E._Lawrence
5-1	"We continue to...": Colonel Robert McClure's Comments to Brigadier General William K. Liebel. Alfred H. Paddock, Jr., *Psychological and Unconventional Warfare, 1941–1952: Origins of a "Special Warfare" Capability for the United States Army* (U.S. Army War College, 1979) 202
6-1	"The strategy is...": Li Tso-Peng, *Strategy: One Against Ten, Tactics: Ten Against One: An Exposition of Comrade Mao Tse-tung's Thinking on the Strategy and Tactics of People's War* (Beijing: Foreign Languages Press, 1966).
7-1	"The line between...": Sun Tzu. [Online] Available: http://www.au.af.mil/au/awc/awcgate/navy/log_quotes_navsup.pdf
7-1	"The *modus operandi*...": Marco J. Caraccia, *Guerrilla Logistics* (U. S. Army War College, 1966) 4.

This page intentionally left blank.

Glossary

SECTION I – ACRONYMS AND ABBREVIATIONS

AA	administrative team A (obsolete)
AB	administrative team B (obsolete)
ADP	Army doctrine publication
ADRP	Army doctrine reference publication
AOB	advanced operations base
AR	Army regulation
ARSOF	Army special operations forces
ATP	Army techniques publication
CA	Civil Affairs
CAO	Civil Affairs operations
CBRNE	chemical, biological, radiological, nuclear, and high-yeild explosives
CJSOTF	combined joint special operations task force
DA Form	Department of the Army form
DD Form	Department of Defense form
DOD	Department of Defense
FA	force area team A (obsolete)
FB	force area team B (obsolete)
FC	force area team C (obsolete)
FD	force area team D (obsolete)
FID	foreign internal defense
FM	field manual
GCC	geographic combatant commander
HHC	headquarters and headquarters company
HHD	headquarters and headquarters detachment
HN	host nation
JP	joint publication
JSOTF	joint special operations task force
JTF	joint task force
MIS	Military Information Support
MISO	Military Information Support Operations
ODA	operational detachment—alpha
ODB	operational detachment—bravo
SF	Special Forces
SFG	Special Forces group
SOCCE	special operations command and control element
SOF	special operations forces

Glossary

SOT-A	special operations team A
SOT-B	special operations team B
SOTF	special operations task force
TC	training circular
TSOC	theater special operations command
U.S.	United States
USAJFKSWCS	United States Army John F. Kennedy Special Warfare Center and School
USASFC	United States Army Special Forces Command
USASOC	United States Army Special Operations Command
USSOCOM	United States Special Operations Command
UW	unconventional warfare

SECTION II – TERMS

Army special operations forces

(DOD) Those Active and Reserve Component Army forces designated by the Secretary of Defense that are specifically organized, trained, and equipped to conduct and support special operations. Also called **ARSOF**. (JP 3-05)

clandestine operation

(DOD) An operation sponsored or conducted by governmental departments or agencies in such a way as to assure secrecy or concealment. A clandestine operation differs from a covert operation in that emphasis is placed on concealment of the operation rather than on concealment of the identity of the sponsor. In special operations, an activity may be both covert and clandestine and may focus equally on operational considerations and intelligence-related activities. (JP 3-05.1)

Civil Affairs

(DOD) Designated Active and Reserve Component forces and units organized, trained, and equipped specifically to conduct civil affairs operations and to support civil-military operations. Also called **CA**. (JP 3-57)

Civil Affairs operations

(DOD) Actions planned, executed, and assessed by civil affairs forces that enhance awareness of and manage the interaction with the civil component of the operational environment; identify and mitigate underlying causes of instability within civil society; or involve the application of functional specialty skills normally the responsibility of civil government. Also called **CAO**. (JP 3-57)

covert operation

(Army) An operation that is so planned and executed as to conceal the identity of or permit plausible denial by the sponsor. (ADRP 1-02)

joint special operations task force

(DOD) A joint task force composed of special operations units from more than one Service, formed to carry out a specific special operation or prosecute special operations in support of a theater campaign or other operations. Also called **JSOTF**. (JP 3-05)

Military Information Support Operations

(DOD) Planned operations to convey selected information and indicators to foreign audiences to influence their emotions, motives, objective reasoning, and ultimately the behavior of foreign governments, organizations, groups, and individuals in a manner favorable to the originator's objectives. Also called **MISO**. (JP 3-13.2)

preparation of the environment
  (DOD) An umbrella term for operations and activities conducted by selectively trained special operations forces to develop an environment for potential future special operations. (JP 3-05)

Special Forces
  (DOD) U.S. Army forces organized, trained, and equipped to conduct special operations with an emphasis on unconventional warfare capabilities. Also called **SF**. (JP 3-05)

special operations task force
  (Army) A temporary or semipermanent grouping of ARSOF units under one commander and formed to carry out a specific operation or a continuing mission. (ADRP 3-05)

theater special operations command
  (DOD) A subordinate unified command established by a combatant commander to plan, coordinate, conduct, and support joint special operations. Also called **TSOC**. (JP 3-05)

This page intentionally left blank.

References

REQUIRED PUBLICATIONS

ADRP 1-02, *Terms and Military Symbols*, 24 September 2013.

JP 1-02, *Department of Defense Dictionary of Military and Associated Terms*, 8 November 2010.

RELATED PUBLICATIONS

These documents contain relevant supplemental information.

Army Publications

Most Army doctrinal publications are available online: http://www.apd.army.mil

ADP 3-0, *Unified Land Operations*, 10 October 2011.

ADP 3-05, *Special Operations*, 31 August 2012.

ADRP 3-05, *Special Operations*, 31 August 2012.

ADRP 4-0, *Sustainment*, 31 July 2012.

ADRP 5-0, *The Operations Process*, 17 May 2012.

AR 40-68, *Clinical Quality Management*, 26 February 2004.

AR 870-5, *Military History: Responsibilities, Policies, and Procedures*, 21 September 2007.

Army Special Regulation 10-250-1, *Organization and Functions, Department of the Army, Office of the Chief of Psychological Warfare, Special Staff*, 22 May 1951 (obsolete).

ATP 3-05.40, *Special Operations Sustainment*, 3 May 2013.

FM 3-05, *Army Special Operations Forces*, 9 January 2014.

FM 3-05.2, *Foreign Internal Defense*, 1 September 2011.

FM 3-05.160, *Army Special Operations Forces Communications Systems*, 15 October 2009.

FM 3-05.203, *(C) Special Forces Direct Action Operations (U)*, 30 January 2009.

FM 3-22, *Army Support to Security Cooperation*, 22 January 2013.

FM 3-24, *Counterinsurgency*, 15 December 2006.

FM 3-76, *Special Operations Aviation*, 28 October 2011.

FM 4-02.43, *Force Health Protection Support for Army Special Operations Forces*, 27 November 2006.

FM 6-05, *CF-SOF Multi-Service Tactics, Techniques, and Procedures for Conventional Forces and Special Operations Forces Integration, Interoperability, and Interdependence*, 13 March 2014.

FM 27-10, *The Law of Land Warfare*, 18 July 1956.

FM 31-21, *Organization and Conduct of Guerilla Warfare*, September 1951 (obsolete).

FM 100-5, *Operations*, 1 July 1976 (obsolete).

ST 31-90-1, *Operations Against Guerrilla Forces*, September 1950 (obsolete).

TC 18-01, *Special Forces Unconventional Warfare*, 28 January 2011.

USASOC Policy Number 32-09, *Operational Funds*, 3 September 2009.

USASOC Regulation 350-1, *Army Special Operations Forces Active Component and Reserve Component Training*, 5 April 2011.

USASOC Regulation 350-12, *Special Operations Forces Mountaineering Operations*, 5 November 2008.

Joint Publications

Most joint publications are available online: http://www.dtic.mil/doctrine/new_pubs/jointpub.htm

Chairman of the Joint Chiefs of Staff Instruction 3500.01G, *Joint Training Policy and Guidance for the Armed Forces of the United States*, 15 March 2012.
http://www.dtic.mil/cjcs_directives/cdata/unlimit/3500_01.pdf

DOD Directive 5105.75, *Department of Defense Operations at U.S. Embassies*, 4 December 2013. http://www.dtic.mil/whs/directives/corres/pdf/520575p.pdf

DOD Financial Management Regulation 7000.14-R, Volume 12, *Special Accounts, Funds, and Programs*, July 2013.
http://comptroller.defense.gov/Portals/45/documents/fmr/current/12/Volume_12.pdf

DOD Instruction 7250.13, *Use of Appropriated Funds for Official Representation Purposes*, 30 June 2009. http://www.dtic.mil/whs/directives/corres/pdf/725013p.pdf

JP 3-0, *Joint Operations*, 11 August 2011.

JP 3-05, *Special Operations*, 18 April 2011.

JP 3-05.1, *Joint Special Operations Task Force Operations*, 26 April 2007.

JP 3-08, *Interorganizational Coordination During Joint Operations*, 24 June 2011.

JP 3-09, *Joint Fire Support*, 30 June 2010.

JP 3-13.2, *Military Information Support Operations*, 7 January 2010.

JP 3-22, *Foreign Internal Defense*, 12 July 2010.

JP 3-24, *Counterinsurgency Operations*, 22 November 2013.

JP 3-26, *Counterterrorism*, 13 November 2009.

JP 3-40, *Combating Weapons of Mass Destruction*, 10 June 2009.

JP 3-57, *Civil-Military Operations*, 11 September 2013.

JP 4-0, *Joint Logistics*, 16 October 2013.

USSOCOM Directive 10-1, *Terms of Reference—Roles, Missions, and Functions of Component Commands*, 15 December 2009.

USSOCOM Directive 350-3, *Joint Combined Exchange Training*, 18 May 2004.

USSOCOM Directive 525-16, *(S/NF) Preparation of the Environment (U)*, 14 November 2013.

Other Publications

Executive Order 9621, *Termination of the Office of Strategic Services and Disposition of its Functions*, 20 September 1945.

U.S. Department of Defense, *Sustaining U.S. Global Leadership: Priorities for 21st Century Defense*, January 2012. http://www.defense.gov/news/Defense_Strategic_Guidance.pdf

Goldwater-Nichols Act (Public Law #99-433), *Department of Defense Reorganization Act of 1986*, 99th Congress, 1 October 1986.

Lodge-Philbin Act (Public Law #597), *An Act to provide for the enlistment of aliens in the Regular Army*, 81st Congress, 30 June 1950.

National Defense Authorization Act for Fiscal Year 2014 (Public Law #113-66), *An Act to authorize appropriations for fiscal year 2014 for military activities of the Department of Defense*, 113th Congress, 26 December 2013. http://www.gpo.gov/fdsys/pkg/PLAW-113publ66/pdf/PLAW-113publ66.pdf

National Security Action Memorandum 57, *Responsibility for Paramilitary Operations*, 28 June 1961. http://www.jfklibrary.org/Asset-Viewer/0HL2ndLn7UapKJRwZ1H2XQ.aspx

National Security Action Memorandum 124, *Establishment of the Special Group Counter-Insurgency*, 18 January 1962.
http://www.jfklibrary.org/Asset-Viewer/qJbe3E_H7kmxvtbyzSb8pw.aspx

Nunn-Cohen Amendment (Public Law #99-661), *An Amendment to Department of Defense Reorganization Act of 1986,* 99th Congress, 14 November 1986.

Taylor, Maxwell D. *The Uncertain Trumpet.* New York: Harper & Brothers, 1960.

U.S. Code, Title 10, Section 127, *Emergency and Extraordinary Expenses*, 2006.
http://www.gpo.gov/fdsys/pkg/USCODE-2006-title10/pdf/USCODE-2006-title10-subtitleA-partI-chap3-sec127.pdf

U.S. Code, Title 10, Section 164, *Commanders of Combatant Commands*, 2011.
http://www.gpo.gov/fdsys/pkg/USCODE-2011-title10/pdf/USCODE-2011-title10-subtitleA-partI-chap6-sec164.pdf

U.S. Code, Title 10, Section 167, *Unified Combatant Command for Special Operations Forces*, 2011.
http://www.gpo.gov/fdsys/pkg/USCODE-2011-title10/pdf/USCODE-2011-title10-subtitleA-partI-chap6-sec167.pdf

U.S. Code, Title 10, Section 168, *Military-to-Military Contacts and Comparable Activities*, 2011.
http://www.gpo.gov/fdsys/pkg/USCODE-2011-title10/pdf/USCODE-2011-title10-subtitleA-partI-chap6-sec168.pdf

U.S. Code, Title 10, Section 401, *Humanitarian and Civic Assistance Provided in Conjunction With Military Operations*, 2011. http://www.gpo.gov/fdsys/pkg/USCODE-2011-title10/pdf/USCODE-2011-title10-subtitleA-partI-chap20-sec401.pdf

U.S. Code, Title 10, Section 402, *Transportation of Humanitarian Relief Supplies to Foreign Countries*, 2011. http://www.gpo.gov/fdsys/pkg/USCODE-2011-title10/pdf/USCODE-2011-title10-subtitleA-partI-chap20-sec402.pdf

U.S. Code, Title 10, Section 2011, *Special Operations Forces: Training With Friendly Foreign Forces*, 2011. http://www.gpo.gov/fdsys/pkg/USCODE-2011-title10/pdf/USCODE-2011-title10-subtitleA-partIII-chap101-sec2011.pdf

U.S. Code, Title 10, Section 2561, *Humanitarian Assistance*, 2011.
http://www.gpo.gov/fdsys/pkg/USCODE-2011-title10/pdf/USCODE-2011-title10-subtitleA-partIV-chap152-sec2561.pdf

U.S. Code, Title 22, *Foreign Relations and Intercourse*, 1994.
http://uscode.house.gov/view.xhtml?path=/prelim@title22&edition=prelim

U.S. Department of Defense, *Quadrennial Defense Review Report*, February 2010.
http://www.defense.gov/qdr/images/QDR_as_of_12Feb10_1000.pdf.

U.S. Department of Defense, *Sustaining U.S. Global Leadership: Priorities for 21st Century Defense*, January 2012. http://www.defense.gov/news/Defense_Strategic_Guidance.pdf

The White House, *The National Security Strategy of the United States of America*, May 2010.
http://www.whitehouse.gov/sites/default/files/rss_viewer/national_security_strategy.pdf

The White House, *The National Defense Authorization Act for Fiscal Year 2014*, December 2013.
http://www.whitehouse.gov/sites/default/files/rss_viewer/national_security_strategy.pdf

PRESCRIBED FORMS

None

REFERENCED FORMS

Unless otherwise indicated, DA forms are available on the Army Publishing Directorate Web site (www.apd.army.mil).

DA Form 581, *Request for Issue and Turn-In of Ammunition*, July 1999.

DA Form 2028, *Recommended Changes to Publications and Blank Forms*, February 1974.

DD Forms are available on the Office of the Secretary of Defense Web site (www.dtic.mil/whs/directives/infomgt/forms/formsprogram.htm).

DD Form 93, *Record of Emergency Data*, January 2008.

Unless otherwise indicated, SGLV Forms are available on the U.S. Department of Veterans Affairs Web site (http://www.benefits.va.gov/INSURANCE/resources-forms.asp)

Form SGLV 8286, *Servicemembers' Group Life Insurance Form Election and Certificate*, May 2009.

Standard Forms are available on the U.S. General Services Administration Web site (www.gsa.gov).

SF 44, *Purchase Order-Invoice-Voucher*, October 1983.

Index

C
counterinsurgency, v, 1-7 through 1-9, 2-3, 2-5, 2-6, 3-1, 3-4, 3-8, 3-10, 3-11, 4-26, 5-24, 6-1, 6-13

counterproliferation, 2-3, 3-4, 3-14, 3-15

counterterrorism, v, 2-3, 2-6, 3-1, 3-4, 3-14, 6-12, 6-13, 6-16, 6-18, 7-20

country team, 2-3, 5-1, 5-3, 5-7, 5-25, 6-12, 6-14, 6-16, 6-17

crisis response force, 4-19, 4-33, 6-18

D
direct action, v, 2-3, 2-6, 2-11, 3-1, 3-4, 3-10, 3-13 through 3-15, 4-5, 4-33, 6-18

F
foreign internal defense, v, vi, 1-8, 2-3, 2-5, 2-6, 3-1, 3-3, 3-4, 3-8 through 3-14, 4-20, 5-24, 5-25, 6-1, 6-11 through 6-17

J
joint combined exchange training, 2-4, 3-9, 5-24, 5-25, 7-4, 7-9, 7-17

P
preparation of the environment, 2-3, 2-4, 3-1, 3-6, 3-15, 4-27, 4-29, 6-2, 7-19, 7-20

S
special operations command and control element, 4-16, 4-18, 5-9, 5-21 through 5-23

special reconnaissance, 2-3, 2-5, 2-6, 2-12, 3-1, 3-4, 3-12, 3-13, 4-5, 5-22, 6-18

special warfare, v, 1-9, 2-5, 3-2, 3-4, 5-5, 6-7

surgical strike, 2-5, 3-2, 3-4

U
U.S. Army John F. Kennedy Special Warfare Center and School, iv, 4-2

U.S. Army Special Forces Command, 3-3, 4-2, 4-3, 4-6, 4-32, 5-9, 5-26, 5-27

U.S. Army Special Operations Command, 3-1, 3-3, 3-4, 4-1, 4-2, 4-30, 4-31, 5-26, 6-19, 7-4, 7-5, 7-6, 7-8, 7-9, 7-11, 7-12, 7-15, 7-17, 7-20

U.S. Special Operations Command, iv, 3-2, 3-3, 3-8, 4-1, 4-3, 4-32, 5-3, 5-8, 5-9, 5-10, 5-24, 5-25, 5-26, 5-27, 6-9, 7-4, 7-6, 7-8, 7-12, 7-19

unconventional warfare, v, vi, 1-1, 1-4 through 1-6, 1-8, 1-9, 2-1 through 2-3, 2-5, 2-6, 3-1, 3-3 through 3-9, 4-20, 4-23, 4-25 through 4-28, 5-14, 5-22, 5-24, 5-25, 5-27, 6-1 through 6-10, 7-14

This page intentionally left blank.

FM 3-18
28 May 2014

By order of the Secretary of the Army:

                          RAYMOND T. ODIERNO
                          General, United States Army
                                  Chief of Staff

Official:

GERALD B. O'KEEFE
Administrative Assistant to the
Secretary of the Army
      1414003

DISTRIBUTION:
Active Army, Army National Guard, and United States Army Reserve: To be distributed in accordance with the initial distribution number (IDN) 115230, requirements for FM 3-18.

This page intentionally left blank.

This page intentionally left blank.

This page intentionally left blank.